Technology Transfer and Management in the Developing Countries: Company Cases and Policy Analyses in Brazil, Kenya, Korea, Peru, and Tanzania

Technology Transfer and Management in the Developing Countries:

Company Cases and Policy Analyses in Brazil, Kenya, Korea, Peru, and Tanzania

by
Harvey W. Wallender III

with the assistance of
Alice S. Boggs
Samuel L. Hayden
Carlos A. Horta
Richard L. Seely
George A. Truitt

This study is the result of a project sponsored and administered by the Fund for Multinational Management Education.

Ballinger Publishing Company • Cambridge, Massachusetts
A Subsidiary of Harper & Row, Publishers, Inc.

This book is printed on recycled paper.

International Standard Book Number: ISBN 0-88410-372-2

Library of Congress Catalog Card Number: 79-12745

Printed in the United States of America

Library of Congress Cataloging in Publication Data

Wallender, Harvey W.
Technology transfer and management in the developing countries.

Bibliography: p.
Includex index.
1. Underdeveloped areas—Technological innovations. 2. Technology transfer. I. Fund for Multinational Management Education. II. Title.
HD45.W33 301.24'3'091724 79-12745
ISBN 0-88410-372-2

to
The Honorable Frank Pace, Jr.
and
Professor Jack N. Behrman

Contents

List of Tables and Figures

Foreword

Enriching life in developing countries through technology transfer is a difficult process. We know how to build an oil refinery in an enclave along the coast, but that is an engineering feat. Far more difficult socially and economically is inducing local firms to produce comfortable, attractive shoes at prices the average citizen can afford.

Technology transfer can help solve this problem. Technology is not magic whereby local managers, knowing the magician's secret formula, can perform production miracles. Rather, production technologies for the vast majority of goods and processes are known—or are knowable. The difficulty is overcoming the hurdles to their implementation.

Harvey Wallender's study takes a penetrating look at the actual process of technology transfer. His conclusions shift the focus from unveiling "secret" technology to improving the capabilities of the local managers who must utilize the technology. The data show that the receiving end of the transfer process is the major obstacle.

Wallender and his associates have compiled and analyzed data from case studies based on sixty-seven consulting projects executed in divergent settings. The conclusions are clear. The payoff in enriched living comes only when local managers assimilate a physical technology *and also* the more general managerial technology necessary for prolonging the benefits of that physical technology. The information provided on stages of user firm development should become a guide for all future attempts at technology transfer.

A sobering inference from these findings is that those technologies that now appear to be most needed are difficult to "transfer" precisely because they are interwoven with cultural attitudes and roles, and with economic structure. As a result, most of the initiative for change must arise from the receiving end of the transfer, that is, from the user firm. This analysis, in looking at the user firm, provides an indispensable perspective on technology transfer and on how that change can be initiated.

William H. Newman
Bronfman Management Scholar
Columbia University
Graduate School of Business

Acknowledgments

It is difficult to acknowledge all dedicated individuals involved in the design, funding, and management of a field research project spanning six countries and involving many firms. David Bell of the Ford Foundation and Frank Pace, president of International Executive Service Corps (IESC), developed the idea for the project in early 1976. A research advisory committee of public and private experts was assembled to guide the preparatory design work. Edward Robison, of the Stanford Research Institute, and Felix Evangelist, formerly of Xerox, worked with the staff of the Fund for Multinational Management Education (FMME) to channel the activities of the research advisory committee. Wallace Elton, Philip Ewald, William Finger, and William Wickman of IESC provided guidance and continuing support.

The actual research design owes much to Professors William Newman, Kirby Warren, David Nadler, and Michael Ginzberg of the Columbia University Graduate School of Business. From the University of North Carolina, Professors Jack Behrman and Joseph Peno contributed greatly to the area of technology policy and development. Several professionals from the Agency for International Development (AID) contributed valuable ideas and concepts to the overall research.

Robert Driscoll of the FMME provided important administrative control and direction and coordinated the various activities over a period of twenty-eight months. Alice Boggs assured the program of an organized and efficient development from design to manuscript.

The success of the field research depended largely on the capabilities of the individuals sent to remote areas. Those who tirelessly collected and analyzed data in the field are cited below:

George Truitt, a development economist and private consultant based in Yorktown Heights, N.Y., generated a variety of cases in Peru and Korea. He also authored both environmental analyses for these countries.

Carlos Horta, Instructor of Finance at Fordham University, wrote the majority of cases on Brazil.

Samuel Hayden, Director of Washington Programming of the Council of the Americas, carried out the case work in Tanzania and wrote the analysis of that country.

Alice Boggs, Project Director at FMME, in addition to her administrative activities, developed cases on Peru and Kenya and wrote the environmental analysis for Kenya.

Richard Seely, Assistant Professor of International Management at Georgetown University, constructed the environmental analysis for Brazil.

Rose Hayden, Deputy Director for Latin America Region of the Peace Corps, and formerly with the American Council of Education, developed special materials relating to the role of education in technology transfer.

Carl Brenden and Ellen Cohen, of Psychometric Consultants, consulted on the application of data processing techniques to the analysis of the program.

These efforts would have been impossible without the financial support of the private sector. The Ford Foundation provided half of the project costs. Their grant was matched by contributions from the Xerox Corporation, Readers Digest Foundation, Allied Chemical Corporation, Pfizer Inc., Continental Oil Company, the Fund for Multinational Management Education, and IESC.

Finally, the assistance of Judith Smith in editing the manuscript and of Laly Garcia Pedroso, Margarita Sans, Ana Maria Crossfield, Halina Zarczynski, and Teri Connelly in typing case studies and the manuscript is gratefully acknowledged.

This project resulted in the contents of the following book, six monographs delivered to various journals and policy forums, and approximately 1,000 pages of case materials made available to IESC for the continued training and development of IESC volunteer activity. The content of these materials is the sole responsibility of the author and does not necessarily represent the opinions of any supporting organization or participating researcher.

The following material is organized for a policy audience in hopes that the findings of this project will assist public and private officials in the creation of more effective programs that improve and expand the role of science and technology.

New York, New York — Harvey W. Wallender III
November 1978

Chapter One

Introduction

"It must be considered that there is nothing more difficult to carry out, nor more doubtful of success, nor more dangerous to handle, than to initiate a new order of things. For the reformer has enemies in all those who profit by the old order, and only lukewarm defenders in all those who would profit by the new order . . . "[1]

Technology transfer lies at the heart of change in organizations and industrial development in the Third World because it is technology that is intimately linked to political and economic independence. The debate over new ideas for strengthening the science and technology communities, particularly in the developing countries, has brought science and technology into a political arena where differing concepts, theories, and orientations confront one another. By focusing on one aspect of this complex issue, this study attempts to contribute to a better understanding of how managerial technologies are received and exploited in the developing countries and argues for a reorientation of international policies and programs designed to improve the overall technological exchange system.

Technology transfer is of great importance to both developed and developing countries. What is at stake is the formation of jobs and of appropriate goods and services and the possible narrowing of the chasm that exists between the have and have not nations of the world. The developing countries are convinced that the control of knowledge by the developed countries leaves them in an ever-increasing state of dependency. For their part, the developed countries have growing concerns regarding the possible loss of competitive capability necessary for economic growth and social development. They are also conscious of the rising militancy in much of the developing world regarding the sovereignty over economic and knowledge resources.

In order adequately to address the fundamental obstacles to technological development, the international debate on science and technology must encompass all aspects of the development and transfer of knowledge. It must consider how knowledge is developed and utilized for production of food systems, for improved health delivery, for industrialization, for exploitation of natural resources, and, ultimately, for the development of psychologically and physiologically healthy individuals. Chapter 2 addresses the debate regarding the conditions under which technology is developed and transferred from the more industrialized societies to the developing world. Since a comprehensive examination of technology transfer demands an understanding of how knowledge is received and utilized in the developing countries, special emphasis will be placed in Chapter 3 on an examination of the role of the developing country productive enterprise, both state and privately owned, and how such an enterprise receives and utilizes technology. Because they have not adequately considered both transferor and recipient characteristics, certain existing policies and programs have failed in their attempt to strengthen the overall science and technology community and to facilitate transfer of technology. Unless the problems and opportunities facing the individual users of technology in the developing world are understood, programs will continue to be launched that may not address the critical bottlenecks that limit the ability of the developing country effectively to seek, acquire, and receive technology.

Increasing political pressures in international discussions have resulted in the call for a new international economic order in which 25 percent of the industrial activity of the world would be in the hands of the developing world by the year 2000.[2] Governments of developing countries have banded together to form specialized petroleum and other commodity cartels to exert influence on the economic and political order. Similar patterns are taking place within intergovernmental institutions such as UNCTAD (United Nations Conference on Trade and Development) in order to establish a more equitable and efficient science and technology environment.

The seventh special session of the United Nations General Assembly agreed to meet in Vienna, Austria, in 1979, in a general conference on science and technology for development. This conference is but one of an ongoing series of major intergovernmental debates that began in earnest in the early 1960s. These conferences continue to reflect concern with establishing a new economic order.

They give special emphasis to identifying strategies to improve the scientific and technological systems that affect development. As part of the ongoing dialogue (specifically in preparation for the 1979 conference), numerous ideas and reports have been generated that argue for greater emphasis on developing the scientific and technological infrastructures of the developing countries.

At the 1979 conference in Vienna, as well as at the fifth UNCTAD conference scheduled for May 1979, there will be a wide discussion on science and technology issues. Many of the countries coming to these conferences will be anxious to learn of the specific options open to them regarding the development of their own technological infrastructures. Research can aid these policy makers in their difficult task by providing them with specific information on how the process of technology transfer works. Though their concern with the international environment for technology sales, transfer, and development is great, their more specific need deals with the day-to-day operation and development of institutions and mechanisms within their own countries in order to facilitate technology transfer. If a developing country understands more about how particular firms function within it, it should be able to identify specific and unique initiatives applicable to its own national setting. In addition, a better understanding of user firms should aid developed countries in designing assistance programs that address the fundamental development problems of the Third World nations.

Unfortunately, as is discussed in Chapter 4 of this book, few current international efforts deal with the limited capabilities in the developing world to receive and use technology. This study attempts to illuminate the problems facing managers in the Third World. It is the belief of this research group that without this and other research efforts, many of the developmental activities now underway will result in little actual change within developing countries. Technology transfer for development is like a game of catch. Much can be done to improve the ability of the pitcher and his toss. However, without a skilled catcher, the game will be incomplete or at best extremely difficult. The analysis in Chapter 3 deals with the catcher in the technology game, the problems he faces, and the opportunities available to improve his capability to receive and utilize technology.

DESIGN AND CONCLUSIONS OF THE STUDY

To contribute to an understanding of the problems of the user firm, particularly as they affect his receipt and use of technology, this research project was organized to study different kinds of organizations in a variety of developing countries. With the assistance of the International Executive Service Corps (IESC), the study team, consisting of five development specialists and academics, analyzed a total of sixty-seven consulting projects with specific organizations from five developing countries—Brazil, Peru, Korea, Tanzania, and Kenya—and prepared detailed case studies based on this analysis. Environmental analyses were developed to describe the political and technological environments of the five countries. An additional survey was carried out covering 405 similar consulting projects with similar technology transfer activities in forty-three developing countries. Through analysis of this information, some of the factors that inhibit the ability of local firms to identify their technological needs and therefore to seek and utilize technology more effectively can be discerned. The important role of the local environment, and how it stimulates or inhibits the local firms in their technology search and utilization, can also be understood.

The major conclusion of this study is that firms of developing countries have a limited capability for diagnosing problems and little understanding of how application of new technologies can be useful and valuable to them. These findings, coupled with the turbulent environment in which these firms operate, explain their weak commitment to planning and setting long-range goals and the consequent reluctance to seek new technologies that could facilitate attaining these goals.

Diagnosis and Planning

Most of the firms examined displayed a weak capability to diagnose their technology needs and to perceive the actual value to their particular firm of any technological improvement or change. For much of the developing world there appears to be little capability for the management activity of planning that leads to improved operations through technological change.

The absence of planning or long-term goal setting by firm managers of developing countries results in a short-term focus—in a concentration on the management activities of organizing and controlling that are necessary for short-term survival. Management in these firms places much greater emphasis on such short-term activities as training personnel in specific systems and instituting certain performance and quality controls than on long-term goal setting or developing systems to motivate personnel.

General and Specific Types of Technologies

The weak planning and leading capability in firms of developing countries suggests the need for general managerial technologies in order for them to be able efficiently to apply specific production or marketing technologies to their operations. If companies do not develop some type of planning capability, they will lack the flexibility and dynamic characteristics necessary to grow and to adapt to changing situations. Therefore, the transfer of systems, machineries, or concepts applicable only in the short term will have less utility than they will when combined with techniques that improve the general managerial capability of the firm. In most of the case studies prepared during the research, the introduction of a specific technology had little long-term effect when it was not complemented by the provision of general managerial systems. These managerial technologies may assist the firm in learning to plan and organize its future needs and developments beyond the immediate problems of improving current operations.

Turbulent Environments

The concept of technology transfer relates to the question of risk taking and change. The case studies and environmental analyses illustrate how difficult it is for local firms to analyze risk and to make long-term commitments for technology change when their local environments are changing rapidly. Uncertainty about changes in governmental regulations or an unstable political environment discourage firms from taking risks and committing to improvements that require long-term change in operations. Contrasting firms in a stable environment such as Korea and in a turbulent environment such as Peru provides an interesting illustration of how the general external environment affects various organizations. Reviewing the environmental impact on the firm also elucidates how specific types of laws and institutions encourage or discourage the proclivity for taking risk and long-term technological planning of individual firms.

There are a variety of other conclusions regarding the applicability of traditional management concepts in the developing world and the role of the effectiveness of different types of change strategies of consultants. These conclusions appear to have less significance in terms of the more fundamental question of the role of management in the technology transfer and development process.

SIGNIFICANCE OF THE STUDY

Assistance programs aimed at improving technology transfer must be specifically targeted in order to use scarce resources most efficiently. This study argues that the development process designed by the management of a particular target organization should be a key consideration in the design of assistance programs. Many current programs, in assuming that the current managerial capabilities of user firms will enable them to take advantage of opportunities for acquiring and developing technology, have not adequately considered user firm managerial processes. What is needed are programs that take into account the important role of the manager in preparing the firm to receive and utilize technology and that thereby contribute to strengthening effective technological infrastructures within the developing countries.

The intergovernmental debate over the transfer of technology has revealed the understandable reluctance of the developed countries to surrender their proprietary technology or to support international initiatives that may restrict research and technology development. Such initiatives, sponsored by the developing countries in international discussions, have been resisted consistently by the developed countries. On the other hand, programs that build local managerial capabilities, improve the technological infrastructure of the Third World, and thereby enlarge global demand for technology will benefit technology suppliers and users. This approach, which can be supported by both developing and developed countries, differs radically from approaches such as international codes of conduct for technology suppliers, which heighten confrontation between developing and developed countries. Chapter 2 highlights the historical conflict between technology suppliers and users; Chapter 3 examines the management behavior of developing country firms in using technology. Based on this examination, Chapter 4 argues for a refocusing of our attention on building local managerial capabilities and local infrastructures in order to reduce the divergence between technology transfer policies of developed and developing countries.

This study attempts through exploratory research to draw general conclusions about technology user behavior as well as to develop descriptive materials that may be used in future research. The case studies and the complementary surveys identify what kinds of factors affect the firms in acquiring and using technology. They were written to depict how different types of firms organize and direct themselves for technological change. In addition, the environmental pieces of Appendix I attempt to show comparisons among the five countries under study in terms of their national, physical, technological, and political structures, and their relationship to science and technology.

IESC AND THE RESEARCH PROCESS

It is not surprising that little research has been done on the behavior of specific firms in the developing countries. Research is costly and it is difficult to apply traditional research methodologies in the very different and distinct environments of the Third World. In addition, it is hard to develop an acceptable data base for examining certain types of research questions or hypotheses. The major goal of this project was to examine how different types of environments, organizational practices, or types of technology influenced successful technology transfer and improvement in a particular firm. Instead of studying different types of technology transfer mechanisms, an attempt was made to learn as much as possible about different types of firms and their particular situations. This required being able directly to observe the managerial behavior of a variety of firms in a number of developing countries.

The International Executive Service Corps (IESC), as an organization committed to transferring technology throughout the world, provided an ideal base for carrying out such a research program. Its general method of technology transfer is relatively constant. It almost always sends volunteer executives as consultants to a firm in a developing country for a two to four month period. IESC is global in scope, operating throughout the developing world in a variety of situations. Because its purpose is linked to changing managerial systems in developing countries, because its format for technology transfer is constant, and because it operates worldwide, IESC is ideally suited to research that attempts to answer questions about what factors affect different firms in the technology transfer process. In addition, IESC has had demonstrable success in its transfer of a wide variety of system specific and general managerial knowledge in the Third World.

The goal of IESC is to assist firms in the developing world in overcoming specific problems that demand specialized know-how on both the managerial and technical levels. IESC is able to provide a variety of types of knowledge at a relatively low cost while avoiding problems of psychological and financial dependence that have hampered many supplier/user relationships. Unlike other sources of technology, IESC's success does not depend on profit generation, market development, or political influence. The primary measures of IESC's success are (1) the degree to which its volunteers can solve problems in the recipient firms; and (2) the degree to which the ability to solve similar problems in the future is transferred from the IESC volunteer to local personnel (a transfer of technology).

IESC was founded in 1964. By 1977 it had carried out over 6,500 individual projects in sixty-five different countries. The original objective of IESC was to establish a practical way to stimulate economic growth in developing nations by aiding local companies in overcoming short-term bottlenecks and gaining access to specific technology which might otherwise not be available. IESC has an extensive recruiting process, maintaining on file over 10,000 volunteers with a variety of managerial and technical industry skills. These volunteers, who are retired executives and technicians willing to share their technical expertise and managerial know-how, can be placed anywhere in the world on relatively short notice. IESC is headquartered in New York and operates with a total permanent staff of sixty-seven and a part-time support staff of twenty-five. The organization itself devotes most of its human resources to the problem of identifying potential clients and locating appropriate volunteers who can meet the needs of those clients. IESC has established a network of country directors and local steering committees in countries where it operates. Country directors are full-time employees of IESC who identify local firms in need of IESC services and aid in the accurate diagnosis of firm problems. The IESC steering committees are made up of local business and government leaders who periodically meet to evaluate the overall IESC contributions to local development. In addition, they act as a vital communication link between the IESC headquarters and the developing countries.

A typical project begins when the IESC country director contacts prospective clients and works with them in analyzing the needs of the firms involved. The description of the needs analysis is sent to New York, where the recruiting department searches for an individual properly qualified for the project at hand. When located, the retired executive is given materials on the client situation and other background information on the national environment of the client firm. Generally within four months of the request from the country director, the volunteer and his spouse are sent to the client's country. The volunteer generally spends two to four months with the client, and leaves with the client various reports and recommendations for improving operations. In some cases, IESC sends a second and even a third volunteer to assist the client in implementing recommendations or maintaining new systems.

IESC focuses primarily on firms with less than $1 million in sales per year (26 percent of projects) and on institutions which generally fall within the areas of education, finance, agricultural promotion, public works, government administration, and hospitals (22 percent of projects completed to date). Firms with annual sales of $1 to 5 million have represented 19 percent of IESC projects; 12 percent have been with companies having sales $5 to 15 million. Firms with over $15 million in annual sales historically make up only 19 percent of IESC's assistance (based on IESC's figures for 1976 and 1977).

To ensure commitment of the local client, a small payment is required; this generally ranges from $4,500 to $13,000 for a three month visit. The fees cover only part of the out-of-pocket expenses of the consultant. Other costs to IESC are covered by grants from U.S. corporations, private foundations, and the United States Agency for International Development.

Through its volunteers, IESC makes available a wide variety of technical and managerial knowledge. This is clearly indicated in the breakdown of functional areas in which IESC has provided assistance:

- Personnel (7 percent of completed programs);
- Financial management (10 percent of completed programs);
- Marketing and distribution (10 percent of completed programs);
- Executive management (36 percent of completed programs);
- Production and design (37 percent of completed programs).

Breakdown of consultations by industry is as follows:

- Agricultural, fisheries, food processing (16.5 percent);
- Construction development and building materials (12.7 percent);
- Machinery, metal products, transportation, and electrical equipment (10.3 percent);
- Textile and apparel manufacturing (9.9 percent);
- Communication, utility, and transportation (8.3 percent);
- Miscellaneous industries and services (6.9 percent);
- Mining and metal refining (5.8 percent);
- Wholesale and retail merchandising (5.6 percent);
- Banking and finance (5.5 percent);
- Paper printing and publishing (5.4 percent);
- Health education and insurance (4.8 percent);
- Petrochemicals and petroleum products (4.5 percent);
- Governmental administration (3.8 percent).

IESC has carried out over 6,500 projects and, given its present organizational capacities, expects to continue at its present rate of approximately 600 projects per year. The majority of IESC's projects have been in the Western Hemisphere, with 30 percent of its programs in South America and 21 percent in Central America and Mexico. East Asia represents another important area, with 26 percent of all projects. The Mideast and Europe (Yugoslavia and Greece) have hosted 879 programs or 19 percent of the total. Africa, an area of growing interest for IESC, has received 133 projects or 4 percent of the total projects.

IESC clearly represents a unique source of technical expertise and managerial know-how for firms and institutions in the developing world. It is capable of transferring industry-, system-, and firm-specific know-how. In addition, where the client demonstrates a continuing need and cannot find an adequate alternative source of ongoing problem-solving technology, IESC can provide such support on a consulting basis.

By helping local firms and institutions solve specific problems, IESC represents a means not only for solution of immediate problems, but for the transfer of technical and managerial know-how from the volunteer to the client and to the environment of the client. As such, IESC represents a core of experience that is valuable when studying the process of transfer of technology.

ORGANIZATION OF THE BOOK

The purpose of the research project on which this book is based was to identify and describe factors that affect local firms in their acquisition and use of managerial technology. The book is organized to move from the present theoretical debate to a more specific questioning of policy initiatives and options in the following manner:

Chapter 2 provides a discussion of prevailing theories and their inconsistencies or applicability to questions of technology and development assistance and national planning.

Chapter 3 summarizes the conclusions of this specific research project on transfer of managerial technology to Third World enterprises.

Chapter 4 indicates policy implications of the research program and suggests the overall need for complementary programs in managerial development.

Descriptions of the general political, economic, and technological infrastructures within the five countries studied are presented in Appendix I. Appendix II describes the research program and a bibliography on research methodologies utilized in this study is also included.

NOTES

1. Nicolo Machiavelli, *The Prince.* London: Oxford University Press, 1968, p. 24.

2. *Lima Declaration and Plan of Action on Industrial Development and Co-operation.* Second General Conference of the United Nations Industrial Development Organization, Lima, Peru, 12-26 March 1975, p. 5.

Chapter Two

International Transfer of Technology: Background to Study

The role of technology in development has been a major concern of developing and developed countries throughout the 1970s. Theories of economic development in the Third World cite technology as a key to rapid industrialization. Although early questions about the contribution of technology to economic development focused on the issue of ownership of productive facilities, the current concern addresses the question of ownership and application of knowledge to the production of goods and services. Increasingly, both the developed and developing countries have realized that one of the keys to development is the ability to obtain, create, or adapt knowledge, and apply it in the productive process. So far, however, most developing countries have put less emphasis on user firm capabilities to receive technology.

EMERGENCE OF TECHNOLOGY AS A CRITICAL INTERNATIONAL ISSUE

Technology was not viewed as a major policy issue for the developed or the developing nations until the last decade. The major assistance programs of the U.S. government in the 1960s, such as the Alliance for Progress, made no reference to technology or scientific infrastructure as a major factor in development, and it was not until the Declaration of the Presidents of the Latin American Nations in Punta del Este in 1967 that the concept of technology was articulated and "Science and Technology were incorporated in the Political Language and Decisions of the Inter-American System."[1]

This articulation in the late 1960s of technology as an important development issue did not lead immediately to a focus on science and technology in policy discussions between the developed and developing countries. In the Conference on the Application of Science and Technology (CACTAL) organized in 1972 by the Organization of American States, the primary focus was the issue of transfer and development of technology.[2] In the years since this meeting, the issue of technology and technological dependency, particularly between the United States and its Latin American neighbors, has been of primary interest to developing and developed countries.

Increasing politicization of the questions about technology—sources, ownership, costs and benefits to developing countries—led to a series of Inter-American working groups in 1974 and 1975. The government ministers participating in these workshops isolated the following issues as bases for discussion:

- Problems stemming from the transfer and development of technology through multinational enterprises;
- The responsibility of host governments to develop local techno-structural environments making technology absorption possible;
- The responsibility of the developed nations to participate with the less developed nations in the development of scientific and technological infrastructures;
- Promotion of institutional mechanisms to improve technology flows to the developing countries.

The identification of these issues is evidence of the ministers' concern with technology suppliers—as sources of technology and their impact on host societies—and with the complexity of the technology transfer process in which governments of developed and developing countries have a responsibility for strengthening technological infrastructures in the Third World.

Actors in the Technology Transfer Process

The four major actors in the technology transfer process, as identified in these intergovernmental meetings and in subsequent literature, are (1) the technology supplier, (2) the environment of the supplier, (3) the technology recipient or user, and (4) the environment of the recipient or user. Transfer of technology can occur only when there is a source supplying the technology and a user receiving the technology. Technology suppliers and users may be public or private profit-making or nonprofit enterprises. The environments of both users and suppliers are conditioned largely by general economic conditions and by the governmental regulations affecting technology transfer. Through the process of transfer, the technology supplier becomes the transferring agent. Figure 2-1 below illustrates the interaction of these actors in the transfer process.

Intergovernmental discussions and research on technology transfer have not focused equal attention on all four actors and have been skewed toward the technology supplier and the technology user environment. Criteria for decisions made by suppliers to transfer technology have been examined in detail, theories on how supplier behavior can impede or accelerate development in the Third World have been posited, and responsibilities of technology supplier (home) governments and user (host) governments in developing local technological capabilities have been articulated. Far less attention, however, has been paid to the characteristics of the technology user. The result is that only one half of the transfer process has been examined. Initiatives undertaken by governments in both the developed and developing countries to improve technology transfer and theories aimed at explaining technology transfer are characterized by a failure to examine adequately the traits of technology users and the conditions that affect their receipt and use of technology. Future initiatives must redress this imbalance and reflect characteristics of all actors in the technology transfer process.

Third World Concerns as Seen in International Discussions

Initiatives of the Third World addressing issues related to the transfer of technology reflect the two concerns cited earlier; that is,

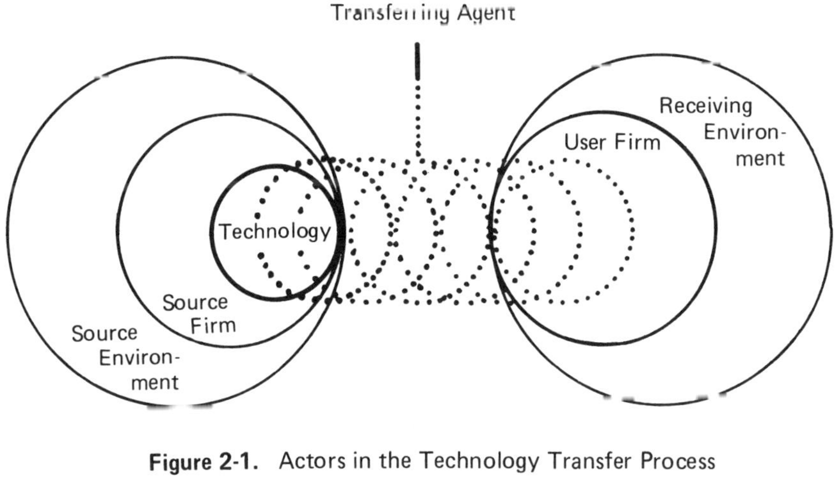

Figure 2-1. Actors in the Technology Transfer Process

behavior of suppliers and the impact of technology transfer on host societies, and the strengthening of developing world capabilities to obtain, receive, and adapt technology.

Behavior of Technology Supplier and Impact on Host Societies. Implicit in this first concern is the control of national governments over the behavior of suppliers in their countries. An early initiative in international discussions to influence this behavior was the Pugwash Conference, held in 1974, at which the notion of a code of conduct for the transfer of technology was aired and adopted by the developing countries. The United Nations Conference on Trade and Development (UNCTAD) has proceeded with negotiations on such a code, which would regulate transfer of technology by multinational corporations (MNCs) and other suppliers. Known as the Group of 77, the coalition of Third World nations active in UNCTAD called for a new technological environment characterized by control over technology suppliers and for careful assessment of the impact of technology transfers and supplier behavior on host societies. This new technological environment was to be structured to accord with the goals of the sixth special session of the General Assembly of the United Nations in May 1974, which resulted in the Declaration and Programme of Action on the Establishment of a New International Economic Order.

Another initiative designed to provide information on the impact of supplier behavior on host societies is the creation of the Centre on Transnational Corporations. At its fifty-seventh session in 1974, the United Nations Economic and Social Council approved a recommendation for establishing a Commission on Transnational Corporations, and, as its operating arm, a Centre on Transnational Corporations (CTC). The centre's work program, drafted in 1975, includes research on the social, economic, legal, and political impacts of transnational corporations on host societies. Priority issues for this research program include the transfer of technology.

Strengthening Developing Country Capabilities to Receive Technology. The United Nations Industrial Development Organization (UNIDO) and UNCTAD have developed special technical service programs designed to aid Third World countries in choosing technologies, bargaining for technology, and developing information systems to support users in their search for technology. Under the assumption that direct government control over the importation and diffusion of technology was the most effective method of improving technology flows into the developing countries, centers were proposed for the study of legislative models to regulate the importation of technology. The design of technical cooperation programs to strengthen the negotiating capabilities of developing countries in negotiating with multinational corporations was also cited as a priority activity of the Centre on Transnational Corporations.

Third World Actions: National Programs

A number of national governments in the Third World have implemented programs that echo the concerns of the Group of 77 in the political debate over the transfer of technology. These programs are aimed at regulating technology supplier behavior in order to increase the flow of technology into their countries while mitigating the perceived ill effects of foreign technology importation and diffusion.

Argentina, Brazil, Mexico, and the Andean Common Market were leaders among developing countries in drafting legislation and imposing new regulations for improved control over the technology transfer process. Some of these regulations reduced the flow of certain technologies into the developing countries. Argentina, Chile, and the Andean Common Market countries were particularly hard hit by this reduction, and governments of some of these countries have revised and softened these regulations. Chile has withdrawn from the Andean Common Market and is now actively seeking new foreign investment. In the wake of its severe economic crisis, Peru is relaxing regulations on foreign investment and is encouraging transfers of needed technology. Argentina, whose regulations were the most stringent, has formulated new policies and instituted new regulations aimed at increasing foreign investment and securing foreign technology.

National programs for improving domestic capabilities in acquiring and using technology vary significantly due to differences among governmental policies and among local environmental and economic factors. The success of Korea in receiving and adapting foreign technology has resulted in part from governmental policies encouraging technology imports by both the private and public sectors, and private/public sector cooperation in all stages of production (from educating and training personnel to research to export of products). The Brazilian government has also supported technological infrastructure development and has, for example, encouraged use by private organizations of public training institutes. Despite continued technology import regulations that are fairly stringent, resulting from current balance of payments problems, Brazil continues to attract new technologies because of its strong infrastructure and market potential.

In contrast to the several issues addressed in varied national programs, the debate at the global or intergovernmental level over technology continues to focus on a uniform approach to controlling technology transfer through regulating suppliers. Initiatives at neither the intergovernmental nor national levels have addressed the specific question of how technology users behave.

Developed Country and U.S. Responses to Third World Concerns

The governments of the developed nations, in response to growing pressure from the less developed countries, launched a discussion on technology and foreign capital parallel to those in the U.N. This discussion, which took place under the auspices of the Organization for Economic Cooperation and Development (OECD), had as its primary purpose the development of international guidelines for the conduct of multinational firms. The guidelines cited the transfer of technology as a major issue. These intergovernmental discussions were designed to demonstrate that the developed nations were also concerned about improving technology transfer, although they hoped that such improvements could occur without promoting rigid and unrealistic theories for global change. In June 1976, the OECD member nations proposed their own code of conduct, which stressed general guidelines for multinational enterprises and identified specific areas for emphasis. These areas included technology transfer and the need to support local research and development, to provide information, and to give due consideration to the problems and goals of the developing nations.

In a policy statement delivered by Secretary of State Kissinger at the Fourth Ministerial Meeting of the United Nations Conference on Trade and Development held in Nairobi in May 1976, the United States addressed the concerns of the developing countries. Major proposals were for the establishment of institutes, centers, and programs that would develop information on the process and impact of technology transfer, incentives for the development and transfer of technology appropriate to the characteristics of specific developing nations, and projects to promote the development of local manpower resources. The U.S. strategy emphasized local industrialization as well as long-term scientific development as an important result of technology transfer. Implicit in this strategy was the U.S. belief that the most efficient manner for dealing with the issue is to focus on the support of local initiatives to expand the capacity of the technostructure to receive and diffuse technology. By focusing on the demand rather than supply of technology, this strategy was viewed as having the advantage of involving the multinational corporation and other suppliers in the technology dialogue, rather than forcing them into reacting after the fact. The strategy differed significantly from that of the Third World, which has concentrated on the supply side of the equation, and attempted to make formal the obligations of technology suppliers.

Several institutions designed to achieve the goals of the U.S. government have been proposed since 1976. An International Industrialization Institute was envisioned as a method of providing expertise to the developing world for long-range planning and feasibility studies. A technology corps was proposed that would provide expertise to technology users through volunteers from the developed world. The U.S. Congress set up an appropriate technology fund which provides financial resources to projects that are labor intensive or that involve marginal sectors. The most recent initiative of the U.S. government is the Institute for Scientific and Technological Cooperation (ISTC), proposed by President Carter in early 1978. The purpose of ISTC is to "expand knowledge and increase availability of technology and skills needed to meet critical problems of people in developing countries."[3] In working with U.S. institutions and institutions of developing countries to improve scientific and technological capabilities in the developing world, ISTC will draw specialists from other U.S. government agencies, universities, U.S. private industry, and the developing countries.

A study completed in 1978 by the National Research Council as background on suggested U.S. initiatives for UNCSTD proposed programs in the two areas of "building capabilities for creating and using industrial technology" and "improving access to the technologies of industrialized countries." Specific recommendations included increased U.S. assistance for creating and strengthening management training institutions and increased assistance to strengthen abilities in the developing world to conduct industrial research, and development of expertise in searching for and selecting technology.[4]

A problem for the U.S. government, however, is that initiatives, particularly the earlier ones, have been undertaken without sufficient consultation with the technology suppliers who work directly with users in the developing world, or without sufficient information on technology user characteristics. According to Nau:

> the technology and development issue arose in such a way that political interests defined what the technology problem was long before the U.S. government or anyone else had reasonable data about the size, direction, absorption, or consequences of technology flows around the world. It is not unusual or even undesirable for problems to emerge by means of a

> clash of political interests. But once this occurs, as it did in the case of early U.S. government discussions with Latin American countries about technology issues, it is imperative to bring to bear as soon as possible hard technical and economic data to suggest the substantive (and not just political) dimensions of the problem. This has yet to happen in U.S. policy discussions of the technology and development problem, let alone in international discussions where politically tainted (at least in the eyes of many U.S. observers, particularly private sector ones) international institutions have sought to assemble and disseminate preliminary data about technology flows.[5]

Without adequate information on the major actors, particularly suppliers and users, in the technology transfer process, formulation and implementation of policies that will be effective are difficult. Programs focusing only on infrastructural development and having designs that do not reflect sufficient consideration of the characteristics of the ultimate users of technology will fail to address critical variables in the technology transfer equation.

Third World Response to Developed Country Initiatives

The responses of the developed countries to the Third World concerns were at first not enthusiastically accepted by developing countries, which continued to focus primarily on the technology supplier and its impact on development. The attempt by the Conference on International Economic Cooperation (CIEC) in 1976 to address the more practical questions of technology transfer was unsuccessful, as the developing countries reiterated that they still consider the major goals outlined by UNCTAD and UNIDO as the prime issues that the developing world should explore for a restructuring of the environment of technology transfer. Policy discussions between developed and developing countries continued to center primarily on codes of conduct, on international mechanisms for altering existing patterns of technology trade, and on ways to develop the technological capacity of developing countries. The developing countries in 1976 did, however, register a growing recognition of the importance of strengthening local technical infrastructures. Statements made at CIEC by the developing countries cited as a major goal the promotion of national capabilities to absorb, adapt, diffuse, and ultimately improve imported technologies. Yet proposed methods for achieving these goals focus primarily on the development of regional or national information and assistance centers and fail, again, to look to the ultimate user of technology and assess his needs.

THEORIES ON AND APPROACHES TO TECHNOLOGY TRANSFER

Concern over the transfer of technology can be understood better if one understands the theoretical background that has fueled the demands of the Third World. This section will review the major theories on and approaches to technology transfer. These focus on the supplier of technology and are

- MNC as oligopoly/monopoly owner of technology theories;
- Dependency theories;
- Screening approaches; and
- Supplier consideration approaches.

In reading the synopses of these theories and approaches, one notes a progression from a highly politicized theory to a more practical approach for explaining technology transfer. The oligopoly/monopoly theories depict the MNC as the owner of most of the world's technology and one that is unwilling to share its know-how. The dependency theories posit that the world is divided into industrialized and nonindustrialized segments, a division that results in the dependence of the latter on the former for growth. Dependency theories isolate technology as a key to development and consequently focus on the MNC as the primary technology supplier. The general conclusion of dependency theories is that MNC presence in the developing world has the negative effect of inhibiting local infrastructure development. The screening approaches hold that MNCs and their technology may be vital to Third World development, but advocate the careful formulation of national technology policies and creation of agencies to screen technologies that will not contribute to national goals. The supplier consideration approaches attempt to describe the practical reasons behind MNC decisions to transfer technology.

Oligopoly/Monopoly Theories

The oligopoly/monopoly theories are based on the assumption that the primary holder of technology is the multinational corporation and that the MNC's long-term objective is to maintain control over its technology. This assumption accounts for the focus of these theories on the MNC as the most important technology supplier. These theories posit that the monopolistic/oligopolistic behavior of MNCs derives from the desire to protect their proprietary knowledge, patents, and highly developed managerial and marketing skills.[6] In applying the general model of oligopoly/monopoly behavior of

MNCs, a number of economists and theorists have focused on various aspects of these practices. One theory emphasizes the "proprietary protectionism" of the multinational corporation in its desire to maintain control over its technology.[7] Other theories stress the "discriminatory monopolist" behavior of MNCs, which results in direct investment in countries with the greatest possibility of profit. Another variation on the general theory posits that MNCs invest abroad in order to exploit new markets after domestic markets for a specific product have been saturated, thereby reaping additional income from a particular technological system.[8]

The significance of the oligopoly/monopoly theories is that they have altered the way in which foreign investment is viewed by policy makers in the developing world. Foreign investment is no longer seen simply as an element of general movement of capital across national borders, but is now viewed with much greater attention paid to the foreign investors (usually multinational corporations) and their impact on developing countries.[9]

Dependency Theories

The theory of dependency was first articulated in the late 1950s through the development of the center-periphery model which divided the world into industrialized centers holding the majority of productive capabilities and periphery areas supplying the centers with raw materials.[10] This model viewed growth in the periphery as a function of growth in the center. Drawing on the oligopoly/monopoly theory, growth in the center was closely tied to the multinational corporation, which was viewed as the primary productive unit because of its proprietary technology and sophisticated management capabilities.

This general dependency theory has taken various forms in the work of other economists and theorists. One version links industrialization and the development of local infrastructures to the presence of MNCs.[11] According to this historical analysis, the local technostructures—consisting of universities, laboratories, and education associations—developed in the industrialized world along with the local productive capabilities. The absence of rapid communications systems and easily available foreign technology resulted in a simultaneous growth in local firms and development of the technostructure in countries such as the United States. The pattern in the Third World has differed significantly because the multinational corporation and other international transfer mechanisms were available to provide technology and technical support by the time the productive sector was beginning to develop. The productive sectors of the developing nations tended consequently to depend on foreign suppliers and failed to create local scientific and technical infrastructures to meet their needs. Dependency on foreign sources of supply through the international transfer of technology resulted, and development of a local technostructure was inhibited.

Another variation on this theory derives from an historical analysis of the growth of the multinational corporation and introduces the notion of a new international division of labor. Under this division, a North Atlantic regime of multinational firms seeks to structure the international economy along hierarchical lines, similar to the internal hierarchical structures of the firms themselves. The result is the geographic centralization of control over the international economy. This theory argues that the occupational distribution of labor, status, authority, consumption patterns, and technology are determined by the center of control and transmitted to the developing countries, creating regional patterns of inequality and dependency. An important element in this system of dependency is the concentration of all R, D, and E (research, development, and engineering) at the center, and the consequent determination by the center of new product introduction and innovation. As an extension of this theory, some note that this concentration of R and D in the home countries not only puts the developing world in a position of "permanent import-dependence for technology," but also deprives the developing country of the multiplier effect that can result from active domestic R and D programs in which local research organizations are in constant touch with local productive sectors.[12] Other theorists note that the presence of foreign investment and imported technical skills often biases the local industrial sector against building up local technical capabilities with the result that a natural form of dependence arises.[13] Foreign investors often, through their bargaining power, discourage or prohibit the use of local talent to reduce risk and protect secrecy of proprietary processes. This theory stresses the issues of the cost of reliance on foreign technology and the inappropriateness of such technology to development planning. The cost of technology is assumed to be too high, resulting from a price set by the monopolistic multinational firms. These theorists advocate building indigenous, innovative capacity in the developing world, with only minimum reliance on the multinational firm in the process.

Screening Approaches

A general conclusion of the dependency theories is that, although technology is needed to achieve development goals, acquisition of technology and the consequent presence of MNCs in developing countries have been detrimental to the strengthening of local productive capabilities and technological infrastructures. Some of the dependency theorists advocate use of local resources for local development in order to minimize interaction with MNCs and thereby to move out of a dependence on MNCs. Other theorists, however, hold that development can be best achieved through use of selected imported technology. These theorists do not discount the significant contribution of MNCs to development, and argue for careful screening of technology and evaluation of its potential contribution to achieving national development goals.

Screening approaches presuppose the formulation and articulation of national policies on technological and developmental criteria for deciding what types and under what conditions technology may be imported. They also call for the creation of national agencies to screen technology projects. Because of their focus on projects dealing with technology and not exclusively on the supplier, the screening approaches represent a more practical and rational approach to development than is promulgated by the dependency theorists.

In screening proposed foreign investment and technology import projects, a developing country should carefully assess the net benefit of the investment.[14] Benefits resulting from new technologies, taxable earnings of the foreign investor, and interaction with local suppliers or laboratories should be weighed against the negative impact of the foreign investment, which might include continued dependence on MNCs for technology and loss of foreign exchange. Through examination of the positive and negative impacts of individual proposals, this approach allows for a more detailed evaluation of the impact of the technology import and may facilitate decision making by the national screening agency.

Supporters of screening approaches generally favor MNC presence through joint ventures and licensing agreements instead of through wholly owned subsidiary arrangements.[15] They also recognize that effective screening of technology projects is contingent upon an understanding of the local market conditions and the characteristics of the local technological infrastructure. Another important element in the screening approaches is the ability of the receiving environment to absorb imported technology and the recognition of the importance of developing local capabilities to use imported technology.

Supplier Consideration Approaches

Supplier consideration approaches represent a fourth type of explanation that focuses on the supplier. One illustration of these models is a 1976 study by Teece[16] which responds to a problem identified in a 1970 UNCTAD report.[17] This report states that it is not clear that there is any relationship between the costs of technology transfer and the type of industry or level of development of the environment of the receiving organizations and suggests that "an identification of the costs (of technology transfer) and the factors which influence them would furnish a meaningful basis for the consideration of national and international policies to reduce them." Teece responds that any attempt to encourage technology transfer by multinational corporations should recognize factors that influence the costs of transferring various types of technologies to various types of users. In examining the total costs incurred by, in this case, the multinational manufacturing corporation in transferring technology internationally, the study isolated four types of costs:

- Transferor characteristics, including experience in international technology transfer and experience in the specific host country;
- Technology characteristics, including transferor's experience with the specific technology, and the number of start-ups by the transferor, nature of technology as part of a manufacturing system, age of the technology, and extent to which other firms are using it;
- Transferee characteristics, including its link to the transferor and its technical and managerial capabilities, and number of years of manufacturing experience;
- Host country characteristics, including utilities, availability of supplies, and skills.

The findings of the study are that the cost of international technology transfers are inversely related to the number of previous start-ups by the transferor, the number of years of manufacturing experience in the user organization, and the age of the technology. Transfer from a parent to a subsidiary is generally less expensive due to the familiarity of each with the operations of the other. Marketing of technology to inexperienced users who may not readily recognize the value of the technology can add considerable cost to a technology transfer. A conclusion from the study is that suppliers prefer to transfer technology to experienced recipients because of the high costs incurred in transferring to inexperienced or ill-trained users.

In a project conducted for the Office of the U.S. Coordinator for the U.N. Conference on Science and Technology for Development, the sponsoring organizations collected information on the views and experiences of 120 multinational corporations towards technology transfer. The following observation from this 1978 project provides insight into the motives of MNCs in transferring technology: "U.S. firms transfer technology whenever there is an opportunity to carry out a commercially viable operation. Companies will consider possible transfer in light of their own particular short- and long-term business interests, organizational formats, transfer strategies, and assessments of the risks involved."[18]

The report goes on to state that a major responsibility of MNCs involved in technology transfer is more effectively to "measure and explain the costs and benefits of their technologies in relation to development efforts."[19]

The supplier consideration approaches reflect a more practical and objective approach to viewing international transfer of technology and represent a digression from the pattern of more politicized theories reviewed above. However, in their focus on the supplier of technology, they, too, fail to address sufficiently those characteristics of the user firm and of the technological infrastructure that affect receipt and use of technology.

CONCLUSIONS ABOUT THEORIES ON TECHNOLOGY TRANSFER

General Conclusions

Differences between the four types of explanations presented in this chapter derive primarily from differences in generality and in ideological base. The oligopoly/monopoly and dependence theories are the most general, and assume that most multinational corporations behave as oligopolists or monopolists, and that MNC presence in the developing world has in most cases impeded local growth and local capabilities to create or adapt technology. This has led to dependence on the MNC as the major supplier of technology. The screening approaches reflect a more practical approach for the developing countries in acquiring and using foreign technology because their premise is that development goals may be most efficiently achieved by using foreign technology in carefully screened and monitored projects. Screening theories presuppose that national developmental goals have been articulated and that agencies established to screen proposed foreign investment projects are qualified to evaluate their potential contribution to developmental goals. Screening agencies have indeed been created in a number of developing countries. However, in some cases, governments have not formulated clear policies or initiated programs for development. The result is that these agencies have no clear guidelines for screening proposed foreign investment or technology import projects. The supplier consideration approaches assume that any explanation of technology transfer patterns must consider the processes through which multinational corporations decide to transfer technology. Their focus on costs incurred by MNCs and incentives and constraints emanating from home and host country regulations represents the most practical approach to understanding MNC considerations in the transfer of technology.

Despite these differences in generality and ideological base, the four types of explanations are similar in their focus on the multinational corporation as the supplier of technology. These theories neglect the other technology suppliers, which are in many cases better equipped to provide the type of assistance the user needs. These theories tend to disregard consulting firms, university laboratories or public training institutions, equipment salesmen, foreign government programs, and private volunteer organizations, all of which are important sources of technology.

The focus on the multinational corporation means that these theories concentrate on the supply of technology. None adequately considers the demand side of the technology equation. The theories are concerned primarily with the multinational corporation as the technology supplier and with the steps a host country can take to regulate MNC operations and the flow of technology into the country. Institutions such as screening agencies and technology information centers have been created for this purpose. What is lacking, however, is a clear understanding of the problems of the ultimate user of technology. As a result, these institutions are less effective than they could be if more information on technology user behavior and needs were available.

Need for Theory on Technology User Problems and Behavior

The theories reviewed above concentrate on the MNC as the technology supplier and on the user environment only as it can be shaped to regulate MNC behavior in the host country. Little attention is paid to the user firm or to the ability of the local infrastructure to provide the local firm with needed technical or managerial assistance. The result has been that the developing countries have poured much energy into the creation of institutions to control technology transfer and very little into programs that address specific needs of user firms.

Technology is not transferred unless it is received. Improving the transfer of technology to the developing world is contingent upon improving user capabilities to receive and utilize technology. Information on the factors that condition receipt by a user organization, however, is sketchy, and understanding of technology receipt and use quite limited. What is lacking is information on how technology users are managed, on what triggers users to seek new technologies, on how users draw on local technology sources for assistance, on how user firms change their management process to improve technology receipt and utilization. This information, once developed, may assist in removing the ideological content of existing theories and current international discussions. The focus must be shifted from the technology supplier to the technology user and concentrated not on the results of technology transfer but on the factors contributing to successful receipt and use of technology by local firms.

DEFINITIONS

Technology

Technology, defined most simply, is knowledge. More elaborately, technology is "any tool or technique, product or process, physical equipment or method of doing or making, by which human capability is extended."[20] Technology used in industrial or business activity may be viewed as

> knowledge necessary for the productive functioning of an enterprise. The term includes process (engineering), management, marketing, and production know-how. It is dynamic, continuing, sequential, and complex. Technology can be found within the public domain, or it can be owned by a government or by private parties.[21]

These definitions indicate clearly that technology is both hardware, consisting of such elements as factories, equipment, and infrastructures, and software, or the nonmaterial components of the productive process, such as education, experience, and organizational structures. Knowledge required to use hardware to produce goods or services is an essential element of technology. This study rests upon a definition of technology that categorizes it in terms of types of knowledge. Such categorization is sufficiently broad to apply to various situations, yet also allows for significant specificity. The following outlines the five knowledge categories:[22]

> *General Knowledge:* that which is publicly available within the society, through books or universities. This is the so-called "common fund" of knowledge.

Industry Specific Knowledge: technical knowledge that is common to an industry and possessed by all firms within it. This is knowledge required for producing a general industry product or managing a general industry process.

System Specific Knowledge: that knowledge and know-how necessary for the production of a specific product. This is acquired through engaging in certain tasks or projects which are linked with the production of a specific item. Other firms producing a similar good would, in the course of actual operation, probably obtain a similar type of knowledge.

Firm Specific Knowledge: production or process knowledge or know-how unique to a specific firm. Such knowledge is not necessarily linked to the production of any given product and may derive from activities peculiar to a single firm. It may include technical knowledge that would go beyond what another firm producing similar products might acquire.

Ongoing Problem-Solving Capability: know-how resulting from experience and necessary to solve production and process problems on a continuing basis.

Managerial Technology

Knowledge for producing goods or delivering services is necessary for the productive functioning of an enterprise. Long-term functioning of the enterprise, however, depends not only on production, process, and marketing knowledge, but also on the know-how necessary, for example, for planning new production processes, organizing resources of the enterprise, and for measuring performance against goals set for the enterprise. These are elements of managerial technology, which is defined as the know-how necessary for managing the productive functioning of an enterprise.

Development in the Third World is in part contingent upon the long-term viability of local enterprises. Importation of technology by a local enterprise is useless unless the organization has the managerial capabilities for planning how the the technology will be used over a long period, for organizing personnel to make most efficient use of the technology, or for diagnosing problems arising from the new technology and seeking solutions to those problems. Strengthening of managerial technologies in local enterprises is thus an important component in Third World development, and is a process that can be facilitated significantly through transfer of managerial skills to local managers. Effective transfer, however, depends on an understanding of current managerial practices of local enterprises as well as on how user organizations can change to manage productive functioning.

Transfer of Technology

Transfer of technology occurs "when technology is transmitted, received, and applied."[2 3] Transfer implies a process linking the technology supplier to the technology user. How technology is transferred depends on the "type of technology and industry, governmental regulatory restraints, the size of the market, the sophistication of the user of the technology, and the costs and benefits to the supplier and user."[24] The "sophistication of the user of the technology," or the strength of management know-how, is particularly critical in the transfer of managerial technology.

Implicit in the definition of transfer of technology is the concept of change in the behavior of the user. An analogy might be the plumber who clears a drain or repairs a damaged pipe in a home. If, in the interaction with the plumber, the homeowner acquires the skills to enable him to clear drains or repair pipes in the future, a transfer of technology has occurred. If, by working with an outside consultant whose assignment is to design an organizational chart for a firm, the manager of the firm learns new techniques that will enable him to reorganize his firm in the future, managerial technology has been transferred. Change in the abilities and behavior of the technology user is the critical element in both of these examples. Transfer of managerial technology is thus defined as a change in the ability to manage the productive functioning of an enterprise.

The concept of transfer of managerial technology is thus intimately linked to the ability to manage technology. However, in order to understand how user organizations manage technology or adjust organizational structures to adapt to new technologies, one must first examine their general management practices, their view of the utility of new technologies, their attitudes towards changing existing management procedures, and willingness to work with an outside consultant to implement new programs for more efficient operations. Existing theories of management can assist in such an examination.

MANAGEMENT THEORY

Understanding how user firms receive and use technology requires an understanding of how they manage their operations. An understanding of major theories on management is necessary to design a program for examining managerial practices in the developing countries that affect the ability of a user firm to receive technology. Specific areas that were considered are:

- How a firm is structured;
- How management principles compare internationally;
- How firms change to improve efficiency; and
- How change in a firm is triggered through an outside consultant.

Structural Characteristics of Management

Major theories relating to the structure of management differ in their focus. Some concentrate on the structure of authority in a firm, others consider the work to be done as the key, while still others focus on human needs and motivation. Contingency theory is based on the assumption that how a firm is managed depends on internal characteristics and external environmental characteristics, and argues for integrating several theories. The development of these theories and the more recent synthesis movements are best explained in terms of their historical evaluation.

Administration and governance theories focus on the concepts of definition and delegation of authority, areas of control of leaders and managers, and a clear definition of goals for the firm and of the responsibilities of individuals.

Henri Fayol and Max Weber[25] argue that organizations function best when rules are clearly stated and communicated, and when individuals understand their individual tasks and receive clear instructions. Individual motivation is associated with economic rewards and recognition of efficient work, and individuals should be advanced according to merit. Thus, examining how an organization functions must concentrate on how goals, job responsibilities, and authority are defined and communicated.

Scientific management theories, often associated with the work of Frederick Taylor,[26] placed even greater importance on planning specific tasks of individuals and creating work environments that support individuals in completing tasks. Careful analysis and engineering studies reveal most efficient methods for performing different jobs. Time and motion studies indicate how plants should be laid out, how tasks should be structured, and what types of individuals are best suited for the tasks. These theories posit that through the scientific analysis of jobs, the most efficient structures and schedules for production can be discerned. Individuals should be paid according to their productivity. As long as goals and responsibilities are clearly communicated, personnel motivation and job satisfaction will increase with efficiencies in the operation of the firm. These scientific management theories stress the disciplines of economics, engineering, and production planning, and view individuals within organizations as rational and economically motivated.

Human relations theorists challenge the view of man as economically motivated and rational. According to these theorists, the success of an organization is closely related to the job satisfaction of personnel and the noneconomic rewards of work. The noneconomic needs of personnel, such as being accepted by others in the firm and feeling that work is contributing to personal growth, are seen as key factors for organizational development. These theories posit that change in organizations for improving operations should focus on developing a more humanistic culture and environment, which will result in more satisfied and fulfilled personnel and therefore increased organizational efficiency.

Advocates of the human relations approach to management criticize bureaucracies and rigid organizational structures. The detailed specification of tasks and formalization of personal relationships within firms contribute to lower productivity and morale. Participation in decision making and work assignments where professional and personal growth is likely are the key factors in the development of the organization. Human relations theorists distinguish between organic firms, where there is less emphasis on functional specialization, precise definitions, and well defined hierarchies, and mechanistic firms, characterized by more rigid hierarchies. Two of these theorists[27] argue further that organic organizations have greater flexibility and can be expected to develop greater technical innovation and capability than can more mechanistic firms.

Contingency theorists have recently begun to resolve some of the differences between the administration, scientific management, and human relations approaches. The contingency theorists argue that internal technologies, or the primary activity of a firm, and the external characteristics of the environment determine whether the human relations, the administrative and governance, the scientific management, or combinations of these theories are most appropriate for an organization. One study found that the technical complexity of the firm conditions what type of management approach is appropriate.[28] Organizations producing goods in mass quantities were found to be most successful when they had mechanistic management systems. Firms with continuous process production systems or where goods are produced in very small quantities (unit production) operated more efficiently with organic systems.

Other theories[29] underscore the importance of the external environment and the firm's internal technology as determinants of appropriate management structures. Those firms with a variety of activities or operating in an environment characterized by frequent fluctuations must have more sophisticated methods for integrating those activities. Firms with less differentiation among departments or units require a less formal structure for integrating the activities of the firm. Contingency theorists thus argue that the frequency of change affecting the firm and the routine or nonroutine nature of its activities are the primary determinants of its management structure.

The concern with the impact of the environment led to the view of an organization as an open system. This view emphasizes, in addition to the internal characteristics of an organization, its relationship to the environment. The success of a firm therefore relates not only to its internal efficiencies but also to its response to the changing external environment.[30] Information on external developments and internal capabilities becomes increasingly important in such an open systems view since the performance of a firm can be fully understood only in relation to its external environment.

The open systems view of a firm and its emphasis on information have led to the argument that the development and proper use of information is the key to effective decision making, and should therefore be the major focus of organizational activities. Much of the prevalent work in operations research and systems analysis addresses the development and use of information and its impact on decision making and on the ability of an organization to survive. Access to accurate information may affect how decision making is structured in a firm.

Management theories offer a variety of approaches for examining organizations, and the diversity in theories has produced a number of issues that may be considered in analyzing the management system of a firm. The most significant of these issues for this study are the following:

- Identification of goals and individual responsibilities;
- Participation in policy development and decision making;
- The organization of work;
- Motivation and its relation to performance;
- The ability of the firm to change and innovate;
- The relationship of the organization to the environment; and
- The decision-making structure.

A study on the transfer of managerial technology to developing country firms must be based on a view of management that not only addresses the structure of management activities, but that also notes how management activities change in response to environmental developments or to innovation. It must also be based on a view of management that allows for comparison among different environments.

Comparative Management Theories

Within the overall debate regarding management theories, comparative management theory has argued the existence of overriding fundamental principles of management. Though many theorists have pointed out the importance of environmental and cultural effects on management,[31] most have agreed that there are certain characteristic functions which provide a general model for describing management activities. Several theorists[32] identified external constraints (educational, sociological, legal, political and economic) and elements of the management process (planning, organizing, staffing, directing, controlling, and policy making) as the fundamental factors conditioning management effectiveness. Others added the concept of management philosophy as an important independent variable.[33] The model was refined further by differentiating between managerial practices and nonmanagerial practices, such as engineering and marketing.[34] These writers argue that there exist underlying principles of management, though there is disagreement on what factors in different environments condition the application of these principles. All of these writers state that the act of goal setting and planning represents the most basic or first step in developing the process of management. Newman, Summer, and Warren[35] simplify the underlying principles by defining the management process as consisting of planning, organizing, controlling, and leading. This model of management is the most applicable when comparing management in developing countries, since it encompasses the significant issues under scrutiny in this study. In addition, the terminology used in this model is widely accepted in both developed and developing countries.

Change Theories

Transfer of managerial technology implies some change in the behavior and capability of an organization. Change theories make it possible to understand how firms in developing countries alter their behavior to adapt to new developments in the external environment, to improve internal efficiencies, and to develop internal capabilities. They are also useful for examining how the behavior of a firm changes as a result of the introduction of a foreign or outside change agent, such as a consultant or a new production technology. A change strategy, and targets of that strategy, are articulated explicitly or implied in any program aimed at modifying, improving, or otherwise altering the behavior of an organization. Change strategies may focus on individuals or on specific management or productive units in a firm, and may use various techniques for effective change.[36]

Individual change strategies focus primarily on values, attitudes, perceptions, and skills of the individual. Data-based strategies involve collecting and feeding data back as the major vehicle for inducing change in an organization. Diagnostic surveys, feedback seminars, and data evaluation groups are examples of mechanisms used in data-based strategies.

An organizational development strategy refers to the attempt to create a different social culture in the working environment of the firm. Emphasis is placed on group and intergroup behavior. Specific targets of change include modifying planning and decision-making processes and altering communication patterns in the firm.

Technostructural strategies are directed at changing the formal structure of the organization or the use of specific technologies by the organization. This type of change strategy is most common among traditional consulting programs.

In addition to analyzing the targets and strategies of change programs, the characteristics of change strategies will be described. A description of change strategies is particularly useful in cases where change results from the intervention of a consultant.

A change program based on acceptance is concerned with helping an organization reduce frustration resulting from a new and uncomfortable situation, and is distinct from strategies that attempt to resolve problems or change situations.

Catalytic strategies characterize those situations where the consultant adds something that has the effect of changing a situation. The consultant may suggest techniques for solving problems. Use of data feedback techniques is common among the catalytic strategies.

Confrontation is a strategy that involves more active participation on the part of the organization. When attitudes toward or perceptions about a situation differ, a consultant may have to force the organization to reconcile these differences before a problem can be solved or improvements made.

Prescriptive strategies are most common in traditional consulting activities and represent the most forceful type of intervention. A consultant using a prescriptive strategy recommends explicit measures and provides explicit instructions to the firm for solving specific problems or changing operating or managerial procedures.

Providing principles and theories equips the firm with guidelines for changing its behavior and is a strategy that offers the firm greater flexibility than do prescriptive strategies. Once supplied with models that it is able to use, a firm can develop an internal capacity for ongoing modification and adaptation of procedures.[37]

The change agent, or, in this study the volunteer consultant, may use a variety of strategies during one project. He may focus on different targets in the firm at different stages of the consultation. One of the research questions addressed in this study is whether and how different targets or strategies in the process of consultation affect the transfer of technology and the resulting change in organizational capabilities.

A standard outline of the stages of a specific change program cites as the first stage the development of a need for change. Other stages are the introduction of a change agent, the diagnosis of specific problems, the examination of alternative courses of action, the selection of a course of action, the production of the change, and the termination of the relationship with the change agent.[38] Of particular importance to the study is the significance of discerning a need to change, the relationship with the change agent, and accurate diagnosis of problems.

MANAGEMENT THEORY AND THE RESEARCH CONSTRUCT

The review in the first part of this chapter of the international discussions on the transfer of technology argues that too little attention has been given to how firms in the developing world receive and utilize technology. What is needed is a clearer understanding of the factors that condition the search of these firms for technology and their subsequent acquisition of technology. Theories that address structural characteristics of management and change in the behavior of an organization are useful in examining factors that affect technology transfer. These theories have been used in the research design of this study in order to address the questions of how enterprises in developing countries are managed, how they view the utility and applicability of technologies to their operations, how developments in their local environments affect their behavior, and how they interact with a change agent in order to absorb and adapt foreign technology.

In addressing these questions, the following categories were isolated as containing those factors that may influence the ability of the firm to acquire and utilize managerial technologies:

Internal Firm Factors
- Comparative Management Characteristics
- Historical Characteristics and Resources

External Factors
- External Environment
- Type of Technology

Process of Consultation Factors
- Diagnosis of Technology Need
- Consultation Program

The data collected in the research all relate to this set of basic factors. In the following chapter are presented the findings resulting from the analysis of the data and illustrate these findings with excerpts from the sixty-seven case studies based on interviews with firm managers that examine these factors in detail. Full case studies are presented as further illustration of the findings and to demonstrate how case study data were organized and compiled. Additional data drawn from the larger survey of 405 cases that were not examined as closely as the sixty-seven detailed cases are also displayed.

NOTES

1. Earl Ingerson and Wayne C. Bragg, eds., *Science, Government, and Industry for Development* (Austin: University of Texas Press, 1976).

2. Robert E. Driscoll and Harvey W. Wallender III, *Technology Transfer and Development: An Historical and Geographic Perspective* (New York: Fund for Multinational Management Education and the Council of the Americas, 1974).

3. ISTC, *Summary and Update: Institute for Scientific Technological Cooperation, Planning Office* (Washington, D.C., October 31, 1978).

4. National Research Council, *U.S. Science and Technology for Development: A Contribution to the U.N. Conference*, prepared by National Academy of Science, National Academy of Engineering, Institute of Medicine (Washington, D.C.: Department of State, 1978), p. 53.

5. Henry R. Nau, "U.S. Policy for Technology and Development: Experience Problems and Private Sector Role," in *Public Policy and Technology Transfer* (New York: Fund for Multinational Management Education, 1978), vol. IV, pp. 43-105.

6. Joseph Peno, "The Multinational Firm: The Utilization and Development of Host Country High-Level Manpower" (Ph.D. Dissertation, Michigan State University, 1975).

Stephen Hymer, *The International Operations of National Firms: A Study of Direct Investment* (Cambridge, Mass.: M.I.T. Press, 1976; originally Ph.D. Dissertation, Massachusetts Institute of Technology, 1960).

7. Ingvar Svennilson, "Technical Assistance: The Transfer of Industrial Know-How to Non-Industrial Countries," in *Economic Development with Special Reference to East Asia*, K. Berill, ed. (New York: St. Martin's Press, 1964).

8. Joseph Peno and Harvey W. Wallender III, "A Contingent Approach to Technology Policy Proposing a Cost/Benefit Analysis," in *Transfer of Technology: The Future of Regulation* (New York: Fund for Multinational Management Education, 1977).

9. Ibid.

10. Raul Prebisch, "Commercial Policies in the Underdeveloped Countries," *American Economic Review* 49, no. 2 (May 1959):251-73.

11. Carlos Martinez Vidal, Speech delivered at a meeting sponsored by the Council of the Americas (Washington, D.C.: February 26, 1979).

12. Thomas N. Gladwin and Ingo Walter, "Technology and Conflict in Multinational Corporate Operations," *Multinationals in Conflict: Lessons in Conflict Management* (New York: John Wiley and Sons, forthcoming late 1979).

13. Charles Cooper, ed., *Science, Technology and Development* (London: Frank Cass, 1973); Keith Pavitt, "The Multinational Enterprise and the Transfer of Technology," in John H. Dunning, ed., *The Multinational Enterprise* (London: Allen & Unwin, 1971).

14. Constantine V. Viatsos, *Inter Country Income Distribution and Transnational Enterprises* (Oxford: Claredon Press, 1974); Constantine V. Viatsos, "Power, Knowledge, and Development Policy: Relations Between Transnational Enterprises and Developing Countries," O.E.C.D. Development Center Conference Paper #3, International Meeting of Researchers on Transfer of Technology by Multinational Firms, 1975.

15. Various screening theories are developed in: M.S. Wionczek, "A Latin American View," in *How Latin America Views the U.S. Investor*, R. Vernon, ed. (New York: Praeger Publishers, 1966); Francisco Sagasti, "A Framework for the Formulation and Implementation of Technology Policies: A Case Study of ITINTEC in Peru," in *Science, Government and Industry for Development*, E. Ingerson and W. Bragg, eds. (Austin: University of Texas Press, 1975), pp. 203-44; Oszaldo Sunkel, "Big Business and 'Dependencia,' A Latin American View," *Foreign Affairs*, April 1972; Santos Mayo, "Interaction between Scientific-Technological and Productive Sectors: Viewpoints of Scientific Research and Education," in *Science, Government, and Industry for Development*, E. Ingerson and W. Bragg, eds. (Austin: University of Texas Press, 1975); and Jorge M. Katz, *Importacion de Technologia, Aprendizaje Local e Industrializacion Dependiente* (Buenos Aires: Instituto Torcauto de Tella, 1972).

16. David Teece, *The Multinational Corporation and the Resource Cost of International Technology Transfer* (Cambridge, Mass.: Ballinger, 1976), p. 100.

17. UNCTAD, "The Transfer of Technology," *Journal of World Trade Law* 4 (September-October 1970):711.

18. Fund for Multinational Management Education, *Public Policy and Technology Transfer* (New York, 1978), vol. 1, p. 9.

19. Ibid., p. 12.

20. Lionel Goldring, "The Management of Science and Technology," Lecture Series at New York University, Graduate School of Business Administration, January-May 1976.

21. Fund for Multinational Management Education, vol. 1, p. 4.

22. Peno, "The Multinational Firm."

23. Fund for Multinational Management Education, vol. 1, p. 4.

24. Ibid., p. 4.

25. Henri Fayol, *Industrial and General Administration*. Translated by J.A. Coubrough (Geneva: International Management Institute, 1929); and Henri Fayol, *General and Industrial Management* (London: Sir Isaac Pitman and Sons, Ltd., 1949); Max Weber, *Social and Economic Organization* (New York: MacMillan Company, 1947).

26. Frederick Taylor, *The Principles of Scientific Management* (New York: Harper and Brothers, 1911).

27. Tom Burns and G.M. Stalker, *Management of Innovation*, 2nd ed. (London: Tavistock Press, 1966).

28. Joan Woodward, *Management and Technology* (London: Her Majesty's Stationery Office, 1958).

29. Paul R. Lawrence and Jay W. Lorsch, *Organization and Environment* (Homewood, Ill.: Richard D. Irwin, Inc., 1969).

30. Daniel Katz and Robert L. Kahn, *The Social Psychology of Organizations* (New York: John Wiley and Sons, Inc., 1966).

31. R.F. Gonzalez and C. McMillan, Jr., "University of American Management Philosophy," *Journal of the Academy of Management* 4, no. 1 (April 1961):33-45; F. Harbison, and C.A. Myers, *Management in the Industrial World* (New York: McGraw-Hill Book Company, 1959).

32. R.N. Farmer and B.M. Richman, "A Model for Research in Comparative Management," *California Management Review* 7, no. 2 (Winter 1964).

33. A.R. Negandhi and S.B. Prasad, *Comparative Management* (New York: Appleton Century Crofts, 1970).

34. H. Koontz, "A Model for Analyzing the Universality and Transferability of Management," *Academy of Management Journal* 12, no. 4 (December 1969):415-30.

35. William H. Newman; Charles E. Summer; and E. Kirby Warren, *The Process of Management*, 4th ed. (Englewood Cliffs, N.J.: Prentice-Hall, Inc., 1977).

36. Harvey A. Hornstein; Noel M. Tichy; and Jay Nisberg, *Organization Diagnosis and Intervention Strategies* (Hopewell Junction: Behavioral Science Associates, 1973).

37. Robert R. Blake and Jane Srygley Mouton, *Building a Dynamic Corporation Through Grid Organizational Development* (Reading, Mass.: Addison-Wesley Publishing Company, 1969).

38. Ronald Lippitt; Jeanne Watson; and Bruce Westley, *The Dynamics of Planned Change* (New York: Harcourt, Brace, and World, 1958).

Chapter Three

Major Factors Affecting Transfer of Technology at the User Firm Level

In many firms in the Third World, a major obstacle to improved technology transfer activities is lack of management capability to diagnose the need for technology. The examination of private and public firms in developing countries reveals that they have limited capability to identify the kinds of technologies they need and, consequently, little impetus for seeking technology. This finding has important policy implications for those institutions in both the developing and developed nations that are working to facilitate technology transfer. Efforts to stimulate research, development, and engineering may be wasted on these firms, which do not have a management base from which to utilize or exploit the resulting technologies. A variety of technology sources are available to developing country firms but are not being used by these firms because of their inadequate managerial resources. Upon examination, it becomes clear that any technology assistance program designed to improve the capabilities of an individual firm must be predicated on an understanding of the managerial capabilities of the firm.

Previous research on technology transfer and development has looked primarily at the supplier firm. A shift of focus to the user firm leads to more precise identification of the specific factors or groups of factors that have the greatest effect on the technology transfer process. It is then possible to take up the question of which of the internal firm or external environment factors are most important in the process, of whether the specific type of technology is important, and of whether or not there are other factors that impede the ability of the firm to adapt and change through the use of technology. All developing country firms, private or public, are engaged in some type of change process. How this process is triggered, how technology is utilized to effect change, and how prepared the firm is to sustain the change are the major questions of this research.

The mechanism used to assess the ability of firms of developing countries to receive and utilize technology was a study of a variety of volunteer consultations undertaken to address specific problems in client firms. When examining a consultation in an individual firm, there must be a differentiation between two possible results: problem solution and technology transfer. Transfer of technology is distinct from problem solution. The plumber who comes to a home and repairs a pipe or cleans a drain has solved a particular problem, but he has not transferred technology. The plumber who comes to a home and teaches the homeowner to fix the plumbing has transferred technology. A successful technology transfer is the transfer of knowledge resulting in change in the capabilities of an organization or individual. This is not to say that projects that solve specific problems and remove bottlenecks affecting the efficiency of an organization are not important. However, there is a significant difference between a problem solution and a true transfer of technology, which involves change as a result of knowledge transferred.

Diffusion of technology, both within the individual firm and in the external environment, is a potential outcome of a consultation that results in technology transfer. Ultimately, diffusion reinforces the transfer of technology by disseminating the knowledge acquired to a larger number of individuals in the firm or to organizations in the local environment.

The focus in this study is on the process of technology transfer because it is more critical to the long-term development and adaptability of the firm than is a given problem solution. Unlike problem solution, which is directed at a specific obstacle, technology transfer effects a change in the capabilities of the firm that enables it to improve its operations.

FACTORS AFFECTING TECHNOLOGY TRANSFER

The variety of factors that influence the ability of a firm to receive and utilize technology can be grouped into three categories. One is the *internal characteristics of the firm*, which include organizational forms, attitudes and motivation of management, manpower capability, and unique task technology. A second category is the *external environment*, including the supporting technostructure of local suppliers, laboratories, research institutes and universities, and the availability of local consultants and skilled manpower. One of the most important external factors conditioning the transfer of technology is the market. Another important element is the government, which through its policies can either impede or encourage effective technology transfer. A third category of factors that influence the receipt and utilization of technology is the *process of consultation*. The strategies and techniques used by the consultant may be an important determinant of whether or not the technology transfer is successful.

For the purposes of this study, these various factors are classified as independent variables, with the results of the consulting project identified as dependent variables:

- Independent variables—internal firm factors (structural characteristics, management capabilities), external factors (local environment, type of technology), process of consultation (diagnosis of technology need, consultation program).
- Dependent variables—problem solution, transfer of technology, diffusion of technology inside and outside of the firm.

Figure 3-1 demonstrates generally how the independent variables interact to affect the outcome of a successful transfer of technology. The possible results of the project—problem solution, transfer of technology, and diffusion of technology—are shown at the bottom as the dependent variables.

The relationship between the independent and dependent variables is complex. For this reason, the study used a number of observation techniques (direct observation, structured interviews, and case studies) to measure the impacts of the different factors on the individual firm and its ability to receive and utilize technology. The IESC consultation program, which uses essentially the same format in every project, provided a common basis for comparing different kinds of transfers, firms, and environments without having to deal with variations between consulting programs of traditional consulting, engineering, or investing firms.

Data Collection

Because of the understandable biases of the volunteer consultant and the client in a project, third party evaluation was determined to be the most accurate measure of the process of consultation and the effect of internal firm and external environmental factors on the

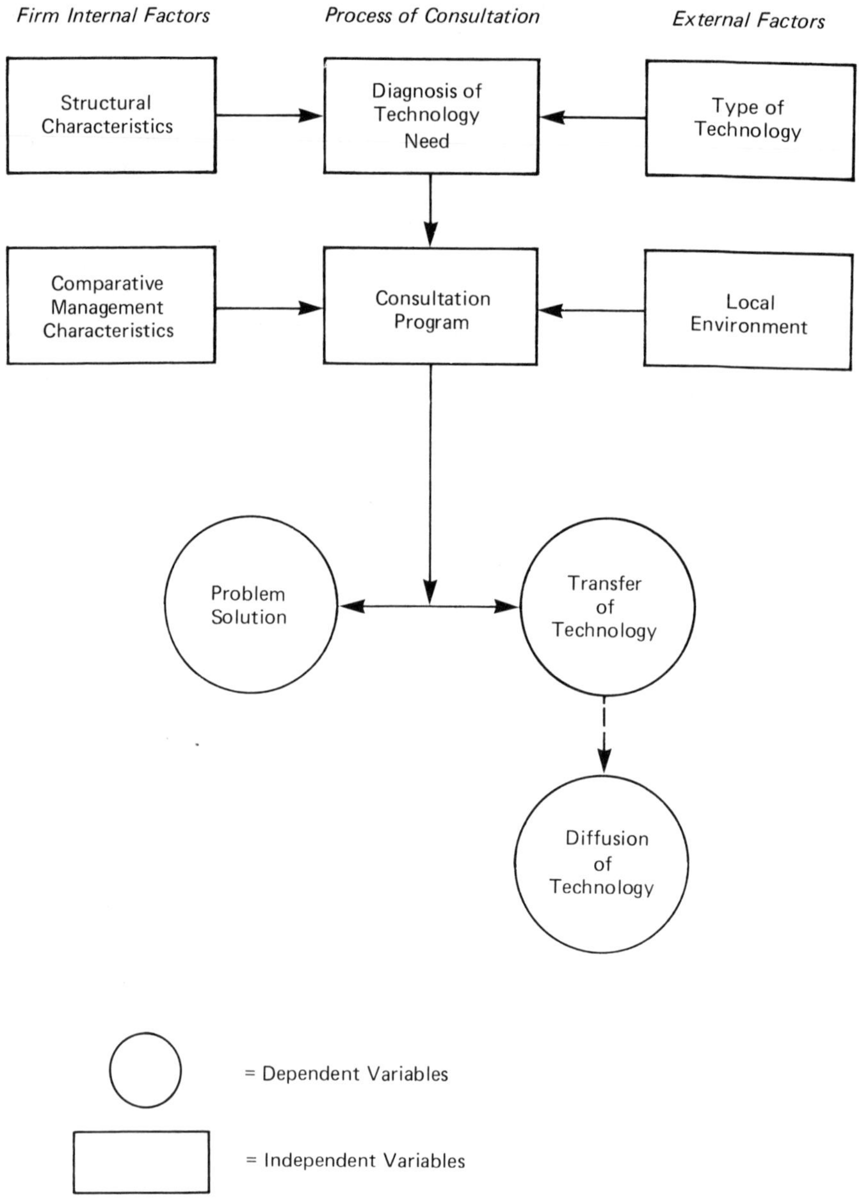

Figure 3-1. Model Relating Independent Variables to Problem Solution, Technology Transfer and Diffusion

results of the consultation. Field researchers visited firms in five selected countries and prepared comprehensive case studies of a total of sixty-seven projects, based on structured interviews with clients, the clients' responses to a prepared questionnaire, and the researchers' review of the project files. The case studies were supplemented by interviews with the volunteer consultants on the projects and by their responses to the same questionnaire administered to the clients. This research process produced for each of the sixty-seven projects a case study and 240 different qualitative measures on the organizational structure and management characteristics of the client firm, the environmental factors affecting the firm, the strategies and technologies used by the volunteer consultant, and the changes that occurred in the firm as a result of the consultation.

Parts of the client-volunteer questionnaire were completed by the field researcher; reliability tests were then applied to the answers given by the volunteers, clients, and field researchers. The results from these tests showed that these three groups of individuals generally agreed in their observations of the same project. In order to generalize the conclusions from the sixty-seven projects, the client-volunteer questionnaire was administered to an additional 405 volunteer consultants on other projects in forty-three countries, yielding a total of 472 volunteer questionnaires.

In reporting the findings from the study, the data collected by the field researchers in their case studies have been relied on more than the data on the larger sample of volunteer responses to the questionnaire. The information derived from the 472 volunteer questionnaires is used primarily to support the findings from the sixty-seven case studies or to illustrate a specific issue that may not have emerged in the individual project analyses.

The case studies of individual projects examine the different internal and external factors conditioning technology transfer and diffusion and the degree and nature of the problem solution and transfer of technology resulting from the volunteer consultations. The primary aim was to measure results from projects in public and private firms, in different industry sectors, and in different environments. Table 3-1 describes the sample of firms by ownership, industry, and country.

Data Analysis

A variety of inferential statistical analysis techniques were used to study the data, including regression analysis, canonical correlation,

Table 3-1. Sample Types of Firms for 67 Case Studies

Number of Projects Examined in Country				
Peru	*Brazil*	*Korea*	*Tanzania*	*Kenya*
15	17	15	12	8
Type of Industry				
Manufacturing	*Service*	*Extractive*	*Continuous Process*	*Other*
22	33	7	3	2
Ownership				
Government	*Private/Family Ownership*	*Private/Public Corporation*	*Joint Venture*	
24	32	6	5	

analysis of variance, and factor analysis. Considering the primary concern of this work for the public policy community, the simpler data comparisons of cross tabulations are used most frequently. The overall regression conclusions are stated briefly on page 47. It should be noted that the author placed greatest analysis confidence in the direct comparisons of the case studies of the field researchers. The information from the case studies provided the primary basis for the study's conclusions. Information from the questionnaires supplemented the case studies and reinforced conclusions drawn from them.

Study Conclusions

There are several important conclusions to be drawn from the direct examination of user firm behavior in acquiring and utilizing managerial technologies. These conclusions relate to the

- Management characteristics of the developing country firm;
- Historical characteristics and resources of the developing country firm;
- Local environment of the developing country firm;
- Type of technology involved in the consulting project; and
- The process of consultation.

The overriding problem of the developing country firm is related to its management characteristics. Most of the firms examined were unable to diagnose effectively the types of problems, or opportunities, confronting them. Without proper diagnosis of technology need or a clear understanding of the value of technological improvements, firms cannot aggressively search for and acquire new technologies.

Certainly, they cannot be expected to respond to national or international technology assistance programs that assume that they are searching for new technologies. The analysis also indicates that those firms that have some experience in technology acquisition, and thus a better understanding of the value of technology, are more likely to seek out new technologies and be receptive to organizational change.

In terms of historical characteristics and resources, most of the firms suffered from a lack of managerial and technical manpower. This is not surprising since the typically small size of local markets and inadequate technological infrastructure have severely hampered, and in some cases precluded, the development of specialized management skills. Many of the firms were not committed to general management development, which was evidenced by their interest in acquiring engineering or specialized production knowledge rather than the more general managerial know-how so essential to planning and diagnosis.

The local environment unquestionably conditions the willingness or capability of the firm to search for and develop new managerial technologies. Most important, the firm must be able to predict the changeability of the external environment so that it can plan for the future. The frequent and unexpected changes in public policy in developing countries affect the risk taking and technological orientation of both public and private institutions. In the face of an unstable political environment—in other words, an unpredictable environment—firms are reluctant to plan for or to make the long-term commitments necessary for technological development. To cite one example, local firms in Peru have been exposed to drastic changes in political and economic policies in the past years.[1] In the early 1970s, government mechanisms were established to encourage the importation of technology. When the country experienced a balance of payments crisis in 1975, the government scrapped many of these incentives and instituted severe import restrictions and currency control mechanisms. These policy fluctuations kept the Peruvian firms in a continual state of confusion. Many firms had planned to buy and install new equipment only to find out later that they could not receive the necessary government assistance to import or convert currency to pay for the equipment.

The most important environmental issue beyond that of environmental turbulence is local or international demand for goods and services. In the study, the firms that were acquiring technology to take advantage of local market opportunities had greater success in terms of technology transfer than those that were responding to internal crises.

The type of technology involved in the consultation program also affected success. In most cases, the consultant thought it essential that the project combine general managerial changes with system specific solutions. Most often the concern of the client firm was to improve production, controlling, or other kinds of specific activities. Yet in dealing with these specific problems, the consultant consistently identified a need for generalized managerial improvements in the area of planning and leading. In most of the cases examined, it was easier to provide solutions to selected problems than to develop a new orientation towards general problem solving, problem identification, and long-term goal setting for the firm.

The process of consultation and the technical capabilities of the consultant obviously affect any particular change program or consultation activity. The evidence indicates that programs combining general management development and system-specific change are more likely to effect a change in the capabilities of the firm and hence result in a successful transfer of technology. When the consultant attempts not only to solve the specific problem but also to change the attitudes of top management and staff towards management technologies, there is greater probability for change in the organization and a true transfer of technology. For the firm to respond positively to long-term change, it is important that the general management problem be addressed not separately but in combination with the system specific problem.

The principal conclusions of the study, all of which promote a clearer understanding of the problems affecting the ability of the user firm to receive, adapt, and utilize technology, can be summarized as follows:

1. Developing country firms lack the general managerial capability to establish long-term goals and match these goals to specific solutions through diagnosis of problems and opportunities. This basic activity is essential for successful technology transfer.
2. Projects addressing both system specific and general management development have the greatest potential for long-term success. Though programs dealing with explicit technology improvements are important to the short-term development of the firm, they must be coupled with general management development if the firm is to improve its capabilities over the long term. Since most firms do not perceive the need for or value of general management development, it falls to the consultant to introduce these techniques along with system specific projects.
3. Few developing country firms comprehend the long-term value of

technological improvement, whatever the type of technology, and thus do not actively seek to acquire technology.

4. Leading (activating) is the most important general managerial characteristic in terms of successful transfers of technology. Top management must understand the importance of long-term goal setting before any fundamental changes in organizing, controlling, and planning can take place.
5. Those firms operating in a turbulent (unknown or rapidly changing) environment are hesitant to make any long-term commitments to technological change. This reluctance can sometimes be offset by a strong market demand or by the presence of a charismatic leader who can motivate personnel and is willing to take the risks required to exploit the opportunities for technological advancement.
6. Those firms that can identify and respond to market opportunities or internal crises will be more aggressive in seeking technology and better prepared to utilize it than those firms that are concerned only with modifying internal operations.
7. The lack of general management resources in the local environment is a major constraint to technology acquisition and utilization. The number of individuals with engineering and production skills far outnumber those with general planning, controlling, and organizing skills. Without these management skills, it is impossible to organize the firm to utilize and exploit technology improvements over the long term.

Thirty-one major factors were identified as having some effect on the ability of the firm to receive and exploit technology. It is to be expected that the abilities of the consultant and the way in which the consultant deals with the client are considered critical to the success of an individual project. Beyond the capabilities of the consultant, the three factors that most clearly determined a successful technology transfer were, in order of importance:

1. Client's ability to perceive the value of technology;
2. Leading (activating) process of management; and
3. Available management resources.

Given an effective consultant, these three factors explain 66 percent of the variance between success and failure in the sixty-seven projects compared. When the management functions of organizing, external technological support, market demand, and technology acquisition history are added, the variance of up to 69 percent can be explained.

These findings suggest that a firm with well-developed managerial resources can overcome many of the internal and external obstacles to technology acquisition and utilization. Given this, what is needed to ensure more effective technology transfers is a clearer understanding of the stage of development of the user firm.

In any given environment, individual firms will be at varying levels of development in terms of their internal capabilities to identify needs, search for technology, and exploit technology opportunities. Some developing country firms may have already developed the internal capability to receive and utilize foreign technology. Their immediate concern should be that of acquiring more information about alternative sources of technology. Most of the firms, however, will not have developed the general management capability to begin seeking much less acquiring or exploiting technology. The critical problem of the user firm in the developing countries, then, is that of developing organizational capability rather than acquiring new information on technology options. For example, the firms studied in Tanzania have fairly ready access to a number of technology sources due largely to the number of assistance programs of European and North American governments. However, few firms are able to exploit these programs because of severe shortages of trained personnel.

It is easier to visualize this situation by thinking of the firm as being in one of eight possible stages of development in terms of technology acquisition and use:

Table 3-2. Stages of User Firm Development for Receipt of Technology

ORGANIZATION DEVELOPMENT	1. Building the initial organizational structure (management and initial technical assistance). 2. Developing an internal problem solving and diagnostic capability at the general management level.
SEARCH AND ACQUISITION	3. Searching for alternative technology after diagnosis and internal problem identification has been carried out. 4. Acquiring alternate technologies. 5. Transferring and exploiting specific technologies.
MAINTENANCE AND MODIFICATION	6. Maintaining and modifying technologies already transferred (product modification and system adaptation).
RD & E	7. Developing unique internal technology capabilities (R&D and product engineering). 8. Exporting (sales) technology to other firms.

These eight stages are cumulative. Any attempt to improve the ability of the firm to acquire alternate technologies at Stage 4 will fail if the firm has not already moved through Stages 1, 2, and 3. Many technical assistance programs, such as the OAS Pilot Project and the United States National Technical Information System, are mismatched to the needs of the user firm precisely because they are directed at a level of development that the firm has yet to attain. Any firm that has not learned how to identify and assess its technology needs will have no basis from which to select among alternative sources of technology.

The results of the study demonstrate that the major problems confronting user firms in the developing countries are associated with the first two stages of development, that is, building a basic management structure and developing the internal capability to diagnose problems and identify the types of technologies that will be of greatest value in improving the capabilities of the firm.

What follows is an analysis of the principal data collected in the study on the internal firm factors (management characteristics and historical characteristics and resources), external factors (local environment and type of technology), and process of consultation. One section is devoted to each of these factors as follows:

Internal: Comparative Management Characteristics
Historical Characteristics and Resources

External: External Environment
Type of Technology

Process of Consultation

Each of these sections includes a comprehensive case study of an individual consulting project in a developing country firm as well as brief references to other projects which exemplify different approaches of these firms to technology acquisition and utilization.

INTERNAL FACTORS: MANAGEMENT CHARACTERISTICS OF THE USER FIRM

To compare management practices among developing country firms in relation to the success or failure of technology transfer, a process of management model was selected that divides management practices into the four elements of planning, organizing, controlling and leading.[2] Planning, a key activity of all managers, entails clarifying objectives, setting goals, establishing policies and methods that will guide the work of the firm in achieving its objectives. Organizing is defined as the assignment of various tasks to different people and the coordination of their efforts. Controlling is the process through which managers measure the day-to-day activities of the firm against the objectives and thereby determine where corrective action is needed. Leading refers to the way in which policies are formulated, formal standards are set, and employees are motivated to execute tasks.

Management theory states that the act of goal setting and planning is the first step in developing the process of management. A simple model in which the planning function is shown to trigger other management activities can be derived as follows.

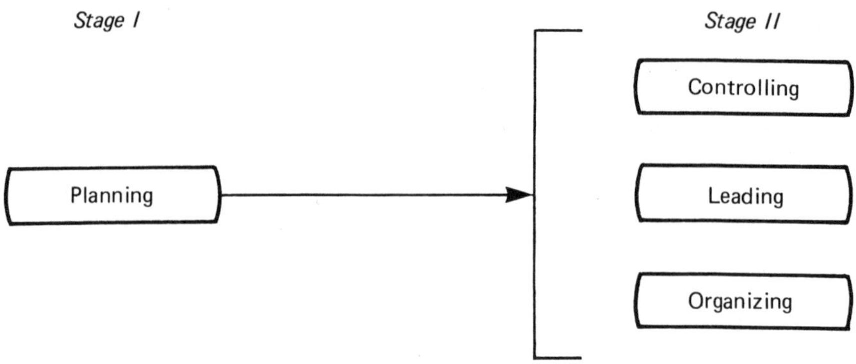

Figure 3-2. Process of Management Development Model I

The traditional interpretation of process of management development does not coincide with the patterns observed in the research in the developing countries. The management principles are no less applicable to the developing country firm than they are to the developed country firm. What the study reveals, however, is a sequence in the process of management development that is strikingly different from that assumed for the developed country firm.

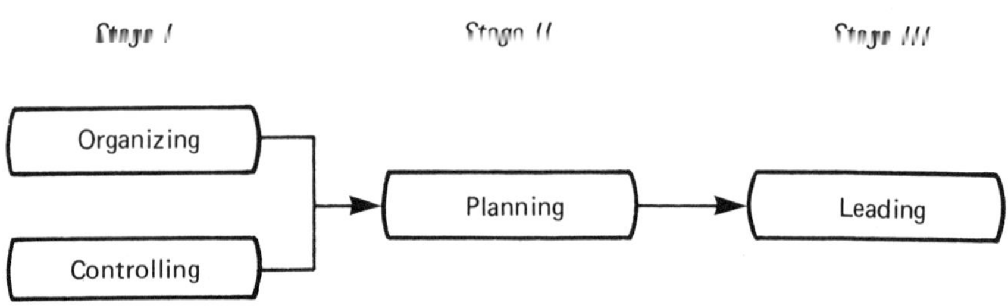

Figure 3-3. Process of Management Development Model II

What this model suggests is that the developing country firm initiates its management process by acquiring a system-specific technology around which it develops its organizing and controlling activities. The technology is often selected intuitively, principally due to a lack of internal capability to assess technological needs and to a short-term view of business opportunities due to uncertainty about the external environment. The planning activity may be initiated later, most often as a result of government regulations or some general belief by management that planning, in an undefined sense, can be useful to the firm. Whether or not planning and goal setting become purposeful, long-term activities of the firm will depend on managerial resources.

Once some form of planning has been initiated in the firm, management may become concerned with developing a more appropriate leadership system through decentralization of decision-making and motivation of personnel. This is difficult to achieve in most developing country firms because they lack the skilled managers who can effectively participate in decision making. It is important here to distinguish between the role of the charismatic leader and the management principle of leading (activating). In the absence of well-developed planning and participatory management systems, it is the individual leader who performs the critical function of setting goals and integrating the activities of the firm.

Planning

Planning is an analytical process that involves three sequential steps: (1) diagnosing the existing situation and problems; (2) defining future goals and objectives; and (3) identifying the kinds of change activities that are required to solve existing problems and achieve the stated objectives. Planning, or defining the transition from the present situation to that envisioned for the future, clearly affects technology transfer in the most fundamental way. On the face of it, meaningful planning takes place in 76 percent of developing country firms, according to the field survey made as part of this study:

Table 3-3. Types of Planning Systems (N = 67)

	Unsophisticated Planning Systems	*Sophisticated Planning Systems*
Number of Firms	16	51
Percent	24	76

The survey brought out this picture in response to questions concerning the planning instruments that are in use in the firms. These instruments were scaled from basic accounting to all-inclusive computerized budgeting. In-depth interviews, however, quickly revealed that the existence of planning activities even in sophisticated form does not mean that management utilizes the planning function in decision making. Most frequently, plans were prepared to fill a government requirement such as providing information on projected demand for foreign exchange or to meet a precondition imposed by a bank before approving a capital investment loan.

The survey of the sixty-seven firms indicates that often plans were prepared by a few persons operating outside of the chain of command, and were seldom consulted by management when decisions had to be made that would affect the future positioning of the company. Ongoing planning in developing country firms tends not to be integrated into management. IESC volunteers who are accustomed to much more functional planning gave low marks to what they saw in the client firms, considering only 38 percent to have a meaningful planning activity.

Table 3-4. Types of Planning Systems (N = 423)

	Unsophisticated Planning Systems	*Sophisticated Planning Systems*
Number of Firms	260	163
Percent	62	38

Most managers indicated that they lacked reliable information in many external phenomena such as population, industry trends, consumer purchasing power, and true inflation, which are relevant to their planning objectives. They live in a setting characterized by a paucity of data. Some data, such as the real prices at which price controlled items are traded, are simply nonexistent; others, such as government statistics on production volumes, are notoriously untrustworthy. Short of skilled planners as they are, managers feel that they have very limited capability to evaluate external data that is available, and therefore incline towards distrusting all of it.

With regard to information on technology, the twin issues of availability and reliability lead to overreliance on foreign sources.

> An example of this reliance on foreign sources of technology is the Remagen Tool Company in Korea. A Korean businessman with no experience in tool production discerned real market potential for

locally produced drill bits. Not knowing where to look in Korea for tool-making machinery, and knowing through an old college friend of a German tool-making plant that was for sale, the owner purchased the plant and transported it to Korea. The owner's assumption that the German equipment would be valuable to him proved incorrect when he attempted to begin organizing production and found that some vital parts were missing, some equipment was rusty, and that, without assistance in repairing and replacing equipment, his purchase would be useless.

Visits to foreign suppliers, foreign consultants, technical journals, and equipment salesmen were cited more frequently than government offices, trade fairs or contacts with competitors as sources of information on technology.

Table 3-5. Sources of Information on Technology Choices (N = 67)

Foreign visits	57	Government offices	33
Foreign consultants	55	Firm visits	30
Journals	49	Trade Fairs	27
Equipment sales	49	Competitor contacts	27
Consultants	45	Foreign missions	19
Manual/texts	42	Newspapers	5

Lack of reliable information on the external economic environment is one reason why most ongoing planning activity is focused on the firm's internal environment and addresses such issues as production schedules or negative cash flow projections for investment projects. Even internal management information systems, however, suffer from shortages of skilled personnel. The case studies from Tanzania are the most illustrative of this problem. All twelve projects from Tanzania suffered from absence of information on internal operations of the enterprises. Had information been available and had there been personnel trained to use such information, the volunteer consultants may have been able to recommend changes with greater confidence in the ability of the firms to implement them.

Another reason—and one that imposes some severe limitation on internal planning as well—is the instability of the external environment. With some effort, reliable information may be obtained on internal firm operations and to some degree on key phenomena in the external environment. It is exceedingly more risky to project into the future. In most developing countries managers must assume that political instability will continue for some time. In a situation where economic policies on taxation, credit, and licensing change frequently, where governments are often replaced by new administrations with different ideological orientations, and where abrupt changes in the system of governance may sweep away legal frameworks, predicting the future can hardly be a science.

> The case of Productos Quemanyo is an example of a firm whose plans for capital improvements were profoundly affected by the requirements of the government. Productos Quemanyo, located in Peru, produces and processes cheese, butter, and yogurt. In response to government tax incentives in the early 1970s to reinvest profits in capital improvements, Productos Quemanyo purchased from foreign suppliers new butter-making equipment, although existing equipment was not in any need of replacement. The old equipment, which was functioning well, is still being stored by the firm to use if the newer machines break down. As part of the capital improvements under this incentive, the firm purchased a yogurt-producing system from a U.S. firm. However, before production of yogurt began, the government issued a new price control that forced the firm to sell yogurt at a price below the production costs. The implementation of the plans of the firm, which were required by the government, adversely affected the operations of the firm as a result of changes in the external environment.

This is a major reason why senior LDC executives devote more of their time to organizing and controlling, and leave planning for later or to a small staff. It seems much more worthwhile to determine where to concentrate the sales force this month than to project demand five years hence.

> Another example of how the external environment can affect the planning process is Turimundo, a hotel design and construction company in Peru, that was established in the early 1970s in response to government incentives aimed at developing the tourism industry. In the first five years, most of the resources of the firm were poured into selecting sites for the seven hotels that were to be built, locating architects for drawing plans for the hotels, and organizing personnel. The deterioration of the Peruvian economy led to frequent changes in regulations affecting Turimundo operations. New import restrictions meant that dishwashers needed for the hotels could no longer be obtained; new taxes levied on travel outside of Peru isolated the client from U.S. hotel

managers and sources of information in the United States he had come to rely on. Land that had been bought and selected as sites for the hotels was subject to a new regulation stating that it could only be used for agricultural purposes. High inflation rates jeopardized the financial base of the company. Because the client had failed to develop long-term or contingency plans to cope with these changes, construction on the hotels had to cease and all energies had to turn towards negotiating with ministry officials or seeking new financing to keep the company afloat. During our interview the client asserted, "I would never had entered this business five years ago had I expected such dramatic changes in governmental regulations." He is not likely to forget the implications of this experience next time when he makes plans for the future.

The lack of planning capability is one reason why developing country firms have such problems defining their need for new technology. Many clients had identified their problems as lying outside the domain of planning, citing very often the need for industry specific engineering technology. Most of the sixty-seven case histories indicate that the workplan of the IESC consultant had to be revised dramatically after his arrival, quite often to focus on changes in planning activities. Being severely hampered in planning by the constraints imposed by the external and internal environments in which they are operating, the managers tend to compensate by seeking immediate improvements in organizing and controlling.

The difficulty with this practice is that to the degree that transfer of technology involves lead time that is longer than the time span that can be confidently predicted, the diagnosis on which it rests tends to be incorrect. Deriving from current organizing and controlling concerns, without taking into account the future internal and external factors, the diagnosis may prove to have been fatally inadequate by the time the technology transfer has been completed.

Organizing

The process of organizing determines how the work of personnel is assembled to achieve whatever goals have been established. How tasks are allocated, how clusters of workers are integrated into a coordinated effort, and how policies are communicated are examples of organizing processes that to some extent condition the ability of the firm to change and adapt to new situations through technology transfer. It might be expected, for example, that those firms that have decentralized decision making and use specialized staffs would have the greatest potential for exploiting knowledge transfers.

What the study shows is that the organizing function is far more developed than is planning. All sixty-seven firms studied had some formal policies although many lacked adequate systems for communicating these policies. Five firms lacked any policy communication system.

Table 3-6. Policy Awareness and Formalization (N = 67)

	Cases	*Percent*
Policies nonexistent	0	0
No formal policy communication and policy unknown to personnel	5	7
No formal policy communication but known to personnel	17	26
Formal policy communication but not known to personnel	5	7
Formal policy communication and known to personnel	40	60

How authority is communicated is a second indicator of the organizing processes of a firm. Job descriptions and organization charts were used in forty-six companies and oral communications in seventeen. The remaining four companies had no visible means of defining authority and personnel responsibilities. Personnel selection was formalized in fifty firms. All but two firms had some type of training program—on-the-job, internal formal programs, formal programs outside the firm, or some combination of these.

In the absence of long-term planning, developing country managers focus on the short term. In many cases, particularly when the firm is new, most energies are poured into organizing resources so as to ensure at least short-term survival. In the developing world, where skilled personnel are scarce, this organizing by itself involves training of some sort. It must involve some communication of general firm policies; and it generally involves a personnel selection system that enables the firm to locate qualified personnel. Hiring personnel from competitors and locating specialized skills in foreign countries were often cited as methods for acquiring needed skills.

> An example illustrating an emphasis on training involves the founder of Mbugua Corporation, Ltd. in Kenya, who had no experience in real estate development when he established his firm. As a result, training of personnel has been a major concern throughout the history of the company, and the policy of the founder, for hiring is to select only those individuals who are

competent in both specific functional areas and in training of younger, inexperienced personnel.

Take another example.

In the early 1960s, Textiles Rimac, a company in Peru, entered an agreement with a leading U.S. textiles firm through which Textiles Rimac received needed technology for a complete reorganization of its production systems. The emphasis in the reorganization was on providing techniques so that the firm could expand production immediately. The resulting transfer of organizing technology was successful—the techniques absorbed in the early 1960s were still being used in 1977 although they were by then outdated. The organizing processes at Textiles Rimac were sufficiently developed to allow absorption and use of foreign techniques. In contrast, the planning processes of the firm were not sufficiently developed to enable the firm to innovate and adapt the 1960s system to changing conditions.

Organizing activity in developing country firm management is usually flexibly defined and stresses on-the-job training in an effort to compensate for rigidities caused by inadequate planning.

Controlling

Controlling is the management function that enables a firm to measure its performance in relation to whatever goals have been established. A majority of the sixty-seven firms studied had some type of control system in place at the time of the consultation. Fifty-seven of the sixty-seven firms had established production standards. Quality control programs existed in forty-four of the firms, and fifty-five had some system of cost controls. Fifty of the firms had instituted budgetary controls.

Participation in controlling systems was significant in some areas. Production standards were set by top management and other staff in forty-one of the fifty-seven firms with established standards; in only sixteen of the firms were standards set by senior management alone. Twenty-one of the forty-four firms with quality control systems had separate quality control departments. There was less staff participation in the formulation of budgetary controls, with thirty-eight of the fifty firms having controls set by senior management only. These findings indicate that the controlling function is generally a routine activity in the developing country firms.

That controlling tends to be more evident than planning is not surprising when we consider the inducements for developing country firms to institute controls. A number of firms in the sample had acquired and installed system specific technologies that required specific quality and production controls in order to be operative. Controls also are often instituted because of competition in the local environment or in export markets that firms may want to penetrate.

> In the case of Omicron in Lima, Peru, the effect of local competition on inducing the firm to institute internal controls can be seen. Omicron produces refrigerators/freezers and air conditioners under a U.S. license. For many years, Omicron operated in a seller's market, as the demand in Peru for refrigerators was well above the aggregate supply, which was restrained by protective tariffs, and the company held approximately 20 percent of the domestic market. In the early 1970s, however, its position began to deteriorate as a result of Andean Common Market regulations which allowed increased imports of refrigerators from Ecuador and Colombia. This new competition forced Omicron to examine its operations and seek assistance in correcting quality problems, such as uneven paint application on its products. In addition, the rapidly rising inflation in Peru at the time of the consulting project induced the firm to institute better cost controls and reorganize production flows to achieve greater efficiencies.

Competition in export markets can also trigger development of controls in developing country firms.

> In Korea, the Yong Dong Motor Company designed a new car. To make a profit on the sizeable investment, it had to increase volume by exporting as well as selling in Korea. In order to begin marketing in the United States and Europe, however, the company had to ensure that its product met safety standards in target countries. A volunteer consultant assisted the firm in identifying specific structural norms and requirements of seat belt and pollution controls that have been established by the U.S. Federal Motor Vehicle Safety Standards. In turn, a number of small businesses supplying components to Yong Dong Motors have been forced to institute more stringent uniformity controls in order to keep their business.

To the firm manager in a developing country, controlling technologies mean essentially two things: first, the need to orchestrate the components of a technological organizing system in a way that assures its reliable functioning; and second, the need to maintain the necessary level of quality and uniformity as determined by competition in the marketplace. This complementary association between controlling and organizing explains the nearly equal attention paid to both the developing country firms.

The relationship between controlling and organizing in developing country firms diverges from the traditional process of management model, in which controlling, as a tool for measuring the implementation of plans, is most closely related to planning. What is found in developing country firms is a focus on organizing and controlling, which may be viewed as the equivalent of short-term planning.

Leading

Leading (activating) is the process by which the manager directs and motivates other personnel to implement the firm's plans. The extent to which the manager responds to the personal needs of employees, resolves conflicts, and establishes rewards bears directly on the performance of the firm. Sometimes the sheer force of charismatic leadership can overcome obstacles to the acquisition and utilization of technology by the firm, but this is distinct from the continuous process of activating, which involves mechanisms for motivating personnel and decentralizing decision making.

The sixty-seven case analyses reveal a variety of decision-making structures among the firms. In forty of the sixty-seven firms, the chief executive developed policy with other management groups or special staff. Production and marketing decisions were made by the chief executive with senior management and special staff in fifty-four of sixty-five firms.

Table 3-7. Decentralization of Decision Making

	Policy Formulation N = 67		*Production Decisions N = 65[a]*		*Marketing Decisions N = 65[a]*	
	Cases	*Percent*	*Cases*	*Percent*	*Cases*	*Percent*
Chief executive	27	40	11	17	11	17
Chief executive and top management	31	46	20	31	29	45
Chief executive, top management and special staff	9	14	34	52	25	38

[a]Only sixty-five responses to this question were obtained.

Fifteen firms had instituted formal programs through which employees can air grievances and make suggestions to the management about firm operations. Thirty-five firms relied on informal means for communicating information from personnel to management or for stimulating morale-building activities.

Another way of looking at the decision-making structures of developing country firms is by the composition of the decision-making group, that is, whether decisions are made by a single individual, by family members who may own the firm, by a process in which many participate in the final decision, or by a hierarchical process. The analysis of this study reveals no definite patterns of decision making in developing country firms.

The absence of long-term plans in the firms studied makes it difficult for managers to use systematic activating processes for directing the implementation of plans. The inchoate state of the activating process seems to have little impact on conditioning the firm to receive technology. This may be because other factors in developing country firms tend to compensate for the absence of the activating management process.

A charismatic leader can play a critical role in motivating the firm and organizing it to acquire technology. The Bosnan Pharmaceutical Company and the Villa Clothing Company, which are presented in full later, are illustrations of companies led by charismatic individuals.

> In contrast to these charismatic leaders is the case in Roberto Torres, S.A., in Lima, Peru. The firm was founded and is owned by Roberto Torres and engages in the import and distribution of hardware and industrial tools. In seeking assistance from the volunteer consultant, Torres outlined the problem areas he discerned, which included marketing, inventory control, purchasing, operating budget, job descriptions, operating manual, new policies on pricing, and long- and short-term planning. What is most interesting about this case is that Torres' diagnosis covered virtually all of the problems of the firm, with the exception of his own inability to trust and motivate personnel, delegate responsibility, and lead. The recommendations offered by the volunteer consultant, many of which implied changes in responsibilities of personnel, have not been heeded. Although implementation of these recommendations would result in more efficient operations for the firm, it would also require a change in Torres' leading style and loss of absolute control of the firm, a change which apparently was unpalatable to Torres.

A second motivating factor that may compensate for the absence of systematic leading is an internal firm environment in which employees feel a strong, almost familylike bond to the firm.

CONRIO, Construces Rio, S.A., is a construction company in Rio de Janeiro, Brazil, that specializes in concrete foundations and structures. Founded in 1961, the company has experienced several rapid growth periods. The number of employees in 1973 was 400, and doubled by 1975. Between 1975 and 1977, the number of employees grew to 1,600. The firm has motivated personnel, even in these periods, largely through programs that reinforce personal loyalty to the firm. For example, the founder of the firm has instituted the custom of holding a luncheon each month in honor of all employees whose birthdays fall within that month. In many other cases, employee loyalty is literally given to the owner family, which plays the role of social protector through such actions as counseling marriages and making personal loans.

The cases of Mwagi and Company and Korean Food Products, which are presented in the following pages, highlight these findings, particularly with respect to the absence of planning, the focus on organizing and controlling activities, and the importance of strong leadership when the planning system of a firm is not well developed.

Case Study: Mwagi and Company (Nairobi, Kenya)

Mwagi and Company (Nairobi, Kenya) illustrates the importance of leadership in motivating firm personnel and in encouraging the development of management practices. The client firm was well prepared to receive the managerial technology needed for better organization and planning because it had diagnosed its situation, identified the technology it needed, and knew why it would be valuable. Although the project focused initially on organizing the firm, the recommendations of the consultant also had a significant impact on the leading and planning functions of the firm. By providing specific organizing techniques, the consultant was able to address the more general activities of leading and planning, the areas where real change in the organization is likely to occur. Because of the forceful personality of the managing partner of the firm, and his commitment to improving the performance of the firm, the project resulted in a significant transfer of managerial technology.

Mwagi and Company was founded in 1969 as an accounting and auditing firm by P.G. Mwagi, who continues today as managing partner of the firm. The firm was run by the founder as a part-time operation until 1970 when he was able to devote full time to the business. In the eight years between the founding of the firm and the visit by the volunteer consultant, the firm grew from a professional staff of four to a professional staff of over 100, six of whom are core management personnel. Turnover rates in personnel have been minimal. At the time of the consulting project, the clients of the firm included a number of Kenya government corporations, and over 60 percent of the small African-owned businesses in Nairobi. By 1977, Mwagi had established itself as the largest African-owned accounting firm in East Africa, and the third largest accounting firm in Kenya. Its services include accounting, auditing, tax preparations, and management advisory services.

Mwagi has a long history of exposure to foreign sources of technology. The founder studied for a number of years in England and worked with the United Nations in Nairobi. The firm is associated with an accounting firm in the United Kingdom, and through this association has ties to accounting firms in some forty countries in Africa, Europe, North and South America, Asia, and Australia. Of the eleven certified accountants at Mwagi, four are from the United Kingdom. Mwagi also recruits certified accountants from other African countries.

There are four identifiable stages in the growth of Mwagi and Company, beginning with the first year of business (1969-1970) when the firm operated on a part-time basis. The second stage, 1970-1972, was troubled in the sense that the company, as the first locally owned accounting firm in Kenya, was looked at askance by other accounting firms and potential clients and had difficulty in locating qualified personnel. Nonetheless, by 1972 the firm had tapped enough of the small business accounts in Kenya to warrant supporting a staff of thirty. In 1973, Mwagi established its first major account, which enabled the firm to expand by recruiting certified accountants both from abroad and in Kenya. Since 1973, the firm has grown rapidly, gaining major accounts from government agencies, and broadening its hold on the small African-owned business market.

As in the case of other locally owned firms that have been founded in Kenya after the country gained its independence in 1963, the major concern of Mwagi and Company has been to establish itself as a legitimate firm; it has faced tough competition from the expatriate firms that have been operating in Kenya for decades. As a result, the management of the firm has had to be aggressive in soliciting accounts, which in the formative years left little time for managing the firm. The firm also has had to compete for scarce qualified personnel in the local environment and has had to locate qualified accountants outside the country. These difficulties in establishing the company as a viable entity, in insuring a strong organizational structure through careful management, and in locating qualified personnel represent the external and internal historical problems of Mwagi and Company.

Internal Firm Characteristics Affecting Technology Transfer. The task technology of Mwagi involves preparing financial statements and tax statements, auditing, and consulting on management issues for a variety of businesses. At the time of the consulting project, the company was organized into departments that included tax, audit, secretarial and registration, and special accounting services (the latter serving small businesses). The heads of these departments reported directly to the managing partner, Mr. Mwagi. While the managing partner consulted with the department heads on major policy decisions, he retained final voice in those decisions. Policies were not formally compiled nor communicated to personnel, but instead were set down in numerous memoranda issued over the years.

The managing partner was responsible for selection of managerial personnel for decisions regarding their promotion. Authority was defined primarily through verbal communications by the managing partner and his staff. Training of personnel, who were recruited from Nairobi and from other countries, was primarily in-house and consisted of both on-the-job training and sporadic formal training seminars. Just prior to the IESC consultation, the firm joined the consortium of accounting firms in Nairobi, which had pooled resources for the purpose of training personnel. The firm also had recruited a full-time training manager from the United Kingdom to set up a formal training program for newly hired accountants.

The managing partner had kept in close contact with his staff since the founding of the firm and continued to be accessible even when the firm grew to over 100 professional staff members. Employee grievances were through the staff and through direct contact with the managing partner. A fund was established by the managing partner from which employees could borrow money in the event of emergencies. The primary motivation for the professional staff, however, was career opportunity: accountants had a far better chance of becoming a partner in a major locally owned firm than would have been the case in a large expatriate firm.

Standards for hiring and performance were set by the managing partner in consultation with department heads. Quality control on Mwagi and Company services was in part guaranteed by the regulations of the accountancy boards or institutes that had registered the accountants. Because of the keen competition with well-established accounting firms, Mwagi was particularly careful that the work performed be well above accepted standards. Budgetary controls were set monthly by the department heads and the managing partner.

The planning area of management was not well developed at the time of the consulting project. Since the founding of the firm, most of the efforts of the managing partner, and those of his staff, had been focused on soliciting new accounts, and, when new accounts were secured, in developing the capabilities and services of the firm so that it could respond appropriately. Company activities consequently evolved in reaction to specific situations rather than in response to a carefully planned development strategy. In the same fashion, the growth in staff reflected growth in the company accounts. It was this lack of control of the organization and planning of the firm that led to the request for a consultation project.

Local Environmental Factors Affecting Technology Transfer. The external environmental factors that have had, and continue to have, the greatest impact on the operations of Mwagi and Company are governmental regulations, the market, and the technological infrastructure. Certainly, in 1969 when the firm was being formed, the risk involved in establishing a locally owned accounting firm was immense. Although the company is now firmly established as a viable business entity, the market continues to be quite competitive, especially in light of the plans of Mwagi for future development. With an increasing number of local businesses as clients, Mwagi has begun to develop a management consulting service designed to teach small business management how to manage their firms and perform preliminary preparation of their accounts for review by Mwagi. While these small accounts comprise the major share of the business for Mwagi, the firm plans to continue to exploit the large accounts market, which at present is still dominated by the expatriate accounting firms. Thus, Mwagi will continue to face stiff competition in its operations.

Government regulations in general support the development of local businesses, which means that Mwagi and Company will eventually be in a very strong position. Specifically, the Kenyanization policy puts Mwagi and Company in a position of being able to attract more readily qualified accountants. Mwagi and Company now holds over 60 percent of the locally owned businesses market. As more and more large enterprises in Kenya become Kenyanized, they may well look to Mwagi as the major locally owned accounting firm, which would put the company in the position of representing much of the business in Kenya.

The local technological infrastructure for accounting in Kenya is inchoate, and Mwagi and Company may support it as much as it supports the company. In the mid 1970s, the Kenyan government created an Institute of Certified Public Accountants, which sets standards for Kenyan accountants and conducts CPA examinations. The institute has been strongly supported by Mwagi and Company, and the firm now requires its professional staff to take the KICPA exams instead of those of overseas professional bodies (primarily Britain and the United States). Until 1977, Mwagi and Company had had very little contact with the other major accounting firms in Kenya. This was due in part to the fact that the firm had to prove itself as a viable business operation before taking an active role in the local accounting environment. In early 1977, Mwagi cosponsored, with five other accounting firms, a training program for accountants, which led to the establishment of solid relationships with the other major accounting firms and the exchange of management ideas.

The physical infrastructure has virtually no impact on the operations of the firm. Change in the industry is relatively slow, and its impact is mitigated by the close tracking of developments in the industry and environment by Mwagi.

Process of Consultation. The market environment in which Mwagi and Company operated, and the difficulties the firm faced in establishing itself as the first major local accounting firm resulted, as noted above, in an almost singular focus on business expansion with little attention on long-term, systematic planning. The managing partner had recognized four years before the consulting project that there would have to be some tightening of the organization, not only to rationalize existing activities but to prepare a solid base for further expansion in Kenya and in other countries through associations with accounting firms. Thus, by the time the client requested the consultation, he had diagnosed the problems of his firm and was committed to solving those problems with the help of an outside advisor. The client specified as the major focus of the consultation assistance: (1) planning the development of the Kenya Institute of Certified Public Accountants and (2) setting up a training program for the firm's staff.

The consultant assigned to the project was a former senior partner and head of the policy group of a large U.S. accounting firm, and was well equipped to address general management and organizational problems at Mwagi in addition to specific problems. He also had broad international experience. After the consultant had reviewed the operations of Mwagi, it was decided that the major focus of the consultation should be on the organization of the firm, with secondary emphasis on the development of the Kenya Institute of CPA. The consultant then began work on preparation of an organizational chart, which would reflect the growth of the company and accommodate continued expansion, and of operating manuals for various departments of the firm. He reviewed procedures in several of the departments and recommended changes in procedures to reduce the work load of top management, thereby allowing time for more careful planning of firm growth.

The primary targets of the volunteer's consultation were the managing partner, the principal accountant (first officer under Mr. Mwagi), and the department heads of audit, special accounting services, tax, and secretarial and registration. Ultimately, the users of the volunteer's recommendations extended beyond that group because of the implementation of suggested procedures that affected the entire staff.

The consultant's strategy for change was both catalytic and prescriptive. In areas where the firm had experience or had a sound procedural precedent, the volunteer acted as a catalyst for the client in developing that precedent into formal policy. In areas where the client had less experience, the volunteer prescribed procedures or techniques for effecting a smoother flow of services. The consultant's information delivery vehicles included general discussion (with the managing partner, the principal accountant, and department heads) of accounting procedures and organization of an accounting firm and explicit models for organizing the client firm and standardizing operations.

Prior to the consultation, Mwagi and Company policy- and decision-making activities centered around the managing partner. Though Mr. Mwagi intended to maintain control over the firm, he also felt that reorganization and redefinition of responsibilities were essential to free some of his own time for planning. The consultant recommended the establishment of a policy-making group headed by the principal accountant and the heads of the four major departments. In addition to their operating duties, the heads of the departments were to have a formal policy-making role instead of the informal role they had held before.

The consultant worked with the operating group manager in preparing a personnel manual to be given to all employees of the company. The manual was compiled from past memoranda, personnel guidelines of the firm with which Mwagi was associated in the United Kingdom, and personnel guidelines from the consultant's accounting firm in the United States. The consultant and the head of the accounting department together prepared an audit manual, which contained standard British and American practices and procedures used by the company. A manual for tax preparation was also written; the secretarial and registration manual was begun but not completed during the stay of the volunteer.

Results of Project. The volunteer consultation addressed organizing as the primary managerial technology. Yet because of the recommendations made for organizing, the leading type of technology played a central role. Planning and controlling activities were addressed as well. The new organizational chart served not only to tighten the organization of the firm, but also to broaden the leading responsibilities of the department heads. By delegating much of the routine work to staff, the new organizational format leaves the managing partner free to formulate plans for the company. The various department manuals delineate performance standards and personnel conduct, and in so doing, address the controlling type of managerial technology.

Because the project was concerned with virtually all departments of the client organization, the potential for diffusion of new techniques was great. The strategy of the consultant seems in fact to have resulted in a significant internal diffusion of managerial techniques. The consequent association by Mwagi and Company with the consultant's accounting firm in the United States may reinforce the managerial techniques learned during the consultation.

In addition to the tangible results of the IESC consultation—new organizational structure; tax, personnel, and audit manuals; association with a U.S. accounting firm—an important intangible result was the relationship that developed between the managing partner and the consultant. The consultant acted as both a sounding board for the managing partner and a catalyst for more effective leadership styles and long-term planning.

Because the project was redesigned after the consultant's arrival to focus primarily on the organization of the firm, there was little opportunity to assist in planning the Kenya Institute of Certified Public Accountants and external diffusion of techniques was minimal.

Review of Major Factors. A number of factors combined in the case of Mwagi and Company to result in successful solution to the problem and in a relatively comprehensive internal diffusion of technology:

- *Client preparedeness.* Mwagi and Company had for several years prior to the consultation recognized the problem with its organizational structure.
- *Charismatic leadership.* Mwagi's strong personality and risk-taking propensity was clear, particularly in his willingness to take on the expatriate accounting community. His leadership was also evidenced by his ability to attract qualified accountants from outside Kenya to work in the first locally owned Kenyan accounting firm.

- *Type of technology.* The primary purpose of the project was to tighten the organization of the client firm. Leading and planning activities were also strengthened by broadening the responsibilities of key personnel in policy and decision making.
- *Nature of firm.* Mwagi has had to be aggressive in seeking clients and willing to take risks to improve operations. The receptivity of the client to the recommendations of the consultant led to immediate implementation of some of the suggestions.
- *Volunteer-qualifications and characteristics.* Not only was the volunteer adept in assessing the needs of Mwagi and Company and in tailoring systems to alleviate the problems of the firm, but he also provided guidance for the managing partner in planning future activities and strategies for company growth.
- *Consultation strategy.* The combined catalytic and prescriptive strategy allowed firm personnel to participate in the solution to the problem where they had experience; where experience and precedent were not strong, solutions through specific recommendations for new procedures were provided.

Several of these factors characterize all consulting projects. The particular combination of factors in the case of Mwagi and Company resulted in a solution to the problem of the client firm, a transfer of techniques, and an internal diffusion of technology.

Case Study: Korean Food Products, Ltd. (Seoul, Korea)

Korean Food Products, Limited requested assistance in solving specific quality control and packaging problems. The consultant chose in addition to address the more general area of planning, and the project consequently resulted in greater awareness by the firm of the importance of long-term plans. Because the planning system of the firm was not well developed at the time of the consultation, the volunteer was able to set up a rudimentary planning system that served to move the firm from a position of reacting to a specific problem to one of planning actively for future growth.

KFP is majority owned by one private individual. It was established in 1966 with 30 percent equity participation by the Korean Development Bank. By 1977, it had paid in capital of about $2 million.

KFP cans fruit, fruit juices and nectars, fish, shellfish, and meats. It produces its own cultured oysters, hogs and mushrooms; but it purchases the fruit from individual farmers and the fish from fishing companies. A KFP subsidiary produces 80 percent of the cans used by the company, with sheet steel imported from Japan. The cost of containers represents some 40 percent of the total value of the product of the company.

The company has production plants in Seoul and in four cities (Taejon, Pohang, Kuryong Po, Chung Mu) in the southeastern part of the country, which is an important agricultural and marine production area. The plants purchase all of their raw materials locally.

About 70 percent of the plant machinery was imported from Japan, with the remaining 30 percent imported from the United States. The equipment vendors provided service engineers for installation and operator training and manuals on processes and maintenance.

The company experienced considerable growth in sales in the period 1973-1977, except in 1974, when real sales declined largely because of the increase in oil prices.

Table 3-8. Growth of Korean Food Products, Ltd.

	1973	*1974*	*1975*	*1976*	*(Est.) 1977*
Sales (in millions of dollars)	7.53	8.00	11.46	15.46	20.62
Increase (percent)		6.24	43.25	34.90	33.38
Wholesale price increase (percent)		42.1	20.2	8.9	10.2
Growth adjusted by wholesale price index (percent)		−35.9	23.1	26.0	23.2

At the beginning of 1976 the company employed 200 permanent employees (excluding 100 in the can-making subsidiary) and 1,500 seasonal workers. By late 1977, it had 350 permanent employees and the number of seasonal workers had increased to some 2,000. During this period, KFP added a new production facility, Cheju Island.

Internal Firm Characteristics Affecting Technology Transfer. The task technology of KFP consists of food processing (cleaning, dressing, cutting, mixing, cooking, purging) and canning (coating, enamelling, washing, filling, sealing, labeling, packing). These tasks are executed in production plants of the company, under the guidance of each plant supervisor. The dispersed small plants are the basic organizational units. The general management in Seoul is headed by the president, who is responsible for key policy decisions but who does not participate in day-to-day decisions. Most decisions are made by the managing director in consultation with the director of trade (marketing) and/or director of production. In addition to these two principal departments, there are the departments of business, general affairs, and research and development.

Company investment in new technology, in addition to being limited by the size of the internal market, has been constrained by governmental policy. After the second five-year plan, which ended in 1971, agroindustries were no longer accorded a high priority, and as a result, KFP access to loan capital for equipment imports no longer was on a par with other industries, which were given higher priority as prime exporters.

The company prepared an annual budget and production and marketing schedules, but did not prepare formal long-range plans. Information on new technology came mainly from visiting equipment salesmen, from technical journals, and from Korean universities. Some information on export markets is provided by the Korean Traders Association.

The quality control mechanisms in the firm at the time of the project were rudimentary, and were largely ineffective. The firm had set standards for specific areas of production in an attempt to protect the market share that it enjoyed. For example, tin-plating standards were set high to provide a wide margin to protect the Eagle brand goods.

External Environment Factors Affecting Technology Transfer. Government policies affect the operations of KFP most significantly through import regulations, allocation of raw materials, and zoning regulations. Government import regulations affect KFP primarily because they permit the entry of only one thickness of Japanese sheet steel for can making, which is not appropriate for all of KFP canning operations. KFP tuna canning operations depend on the amount of tuna fish allocated to the company by the government. The scheduling of government allocations importantly affects productivity and profitability since tuna is harvested during a period of the year when few other raw materials are available. Finally, government zoning regulations have prevented the company from expanding its existing buildings in ways that would be most economical to accommodate the need for additional processing facilities.

Most KFP products are sold on the domestic market. There it competes with a dozen other companies; but having won some 35 percent of the market for its Eagle brand, KFP is the leader in the field.

In view of the relatively short history of processed foods in the Korean consumer market, neither KFP nor its competitors have ventured beyond canning. The companies are more concerned about obtaining a satisfactory return on their investments in canning facilities than about developing new product lines and packaging. In that sense, technological change in the industry, and its effect on the company, has been limited.

Raw materials purchased from small independent producers are often poor in quality and supplies often insufficient. Unreliable delivery of raw materials also causes problems for KFP operations. The physical infrastructure presents no real problems for the firm.

Process of Consultation. Concerned about the rising cost of imported sheet steel, the management of KFP turned to the Korean Traders Association for assistance in determining what would be the least expensive type of container for each company product, consistent with good packaging practice and taking into account the required investment. The management of the firm also hoped to improve the efficiency of the packing lines and solve certain problems with can linings (for example, the strawberry nectar assumed a grey-pink color in storage, and the plum and grape nectars acquired a strong lacquer flavor and odor). The company wanted to evaluate different containers, such as thin wall disposable glass jars, and in general develop a long-range container plan that would permit it to maintain its industry leadership without losing profitability. The firm also hoped to obtain advice on the production and packing processes.

In response to the KFP request for assistance, the Korean Traders Association arranged for an IESC consultation. IESC provided the volunteer services of a retired executive from the H.J. Heinz Company, who had spent a good part of his career in Australia, Venezuela, and Mexico as president of Heinz subsidiaries. The volunteer not only had experience in working in foreign environments, but had previous consulting experience from two IESC assignments and an AID project. Before undertaking the project, the volunteer went to the University of California to obtain information on new packaging techniques, and visited in Tokyo with an executive of a Japanese packaging firm which was using a new pouch type of package (not in use in the United States).

When the volunteer arrived in Seoul, he spent several days with the head of KFP quality control and with the managing director discussing what they believed were the problems of the firm. He then visited each of the six plants of the firm. It was during these visits that the volunteer realized that the real problems were related to management and not simply technical difficulties in the packaging procedures.

The consultant recognized, even more clearly than did company management, that the ability to improve packaging depended to a large extent on the ability to plan ahead. Only through long-range planning could the company make the necessary changes in existing facilities, which were poorly laid out and severely limited in space, to permit economical mechanization and improvements in production methods.

Accordingly, the consultant and management agreed upon a new plan of work, and a schedule of activities for each week was formulated to address the following areas:

- Identification on the options for improving packaging, taking into account market, cost, quality, and methods;
- Development of a strategic plan of facilities, to establish priorities for further investigation;
- Assistance in the planning of the Busan production project (meat and fish company) and the new juice line project; and
- Development of criteria for new product development.

The volunteer worked closely with the head of research and development, the trade manager, and their teams in reviewing information from company headquarters and collecting other information from outside sources. He made return visits to each plant and also visited a glass-making plant, a tinplate mill, a lacquer manufacturer, and several supermarkets. The volunteer recommended changes for KFP based on observations made during visits to firm plants. He also held frequent discussions with the managing director and top management, both in informational meetings and in the weekly staff meetings, in which he actively participated.

At the time of the consulting project, the management of KFP was aware that major decisions about the firm were required to solve its problems with packaging and quality and with overall efficiency. The volunteer thus acted as a catalyst in assisting the firm in its long-range planning activities. His strategy for change, in addition to being catalytic, was prescriptive in that he recommended specific techniques for tin-plating, for new containers, and for new product development. A summary of his recommendations follows:

Containers. The volunteer estimated that $136,000 or 20 percent could be saved by applying thinner tinplate and recommended an experimental program, in collaboration with the can-making subsidiary, to reduce gradually the tin coat thickness. He pointed out that KFP was using thicker plating than any competitor, and that the U.S. Canning Manual recommended a thinner plating.

The volunteer also provided a preliminary evaluation of can

substitutes and recommended further research. This evaluation covered glass jars (domestic), Tetra-Pak (Sweden), the laminated pouches (Japan); and pull-tab cans (Sam Wha Can Company, under U.S. license). The sterilizable, shelf stable pouch that the volunteer had learned about in Japan showed particular promise.

Facilities Planning. Convinced that the company needed to invest in new equipment, but knowing that the existing plants could not accommodate such modernization, the consultant outlined the long-range planning of facilities. He suggested that KFP develop a master plan of facilities; establish a five-year production schedule; and formulate a five-year marketing plan based on a review of internal and external markets and projections on new product and new market opportunities. The consultant believed this type of plan would provide for increased mechanization of canning by introducing new equipment. For information on processing, he suggested that company managers visit the HJH plant at Shepperton, Victoria (Australia), one of the most modern in the world, and that they consider leasing or buying equipment from HJH to increase production efficiency.

The proposed plan also specified criteria for site selection, including total area, water, power, waste disposal, personnel, transportation, raw material supply, packaging material, and supply area zoning.

Busan Production Project. KFP was planning to convert a warehouse building it owned in Busan into a meat and fish canning plant. The job of the consultant was to evaluate the feasibility of the plan. He began the feasibility study by raising economic questions. The competitiveness of the plant depended on the ratio of Korean to Japanese wages, and he questioned whether adequate consideration had been given to the possible changes in this ratio as well as in freight rates.

While the supply of lamb and mutton seemed assured from New Zealand, there was no such guarantee regarding tuna, even though Korea was then harvesting over 100,000 tons of tuna annually. The consultant suggested that KFP plan a buffer stock of frozen tuna. He also pointed out that tuna prices had doubled over the preceding five years and that profitability depended on total recovery in the form of by-products. He suggested that the company bring in a tuna specialist for consultation, since the competition for use of tuna by-products was fierce.

To assess the prospective plant site, the volunteer studied the packing operation at the Kuryong Po plant to determine area and facilities requirements. He saw two advantages in using the Busan warehouse: KFP owned the building and the adjacent land and the docks were nearby. Against this were arrayed several disadvantages: the warehouse was located in a crowded industrial area where competition for labor was keen; the water supply was not reliable, and severe restrictions on the effluent discharge had to be anticipated; it would duplicate the existing KFP plant in Busan; and, most important, there was the problem of fixed dimensions since government regulations prevented further construction on the site. The layout of the building was not sufficiently flexible to absorb production from other plants; the consultant recommended that the company consider other sites, preferably at one of the other plants where expansion would require less investment and would yield quicker profits.

New Juice Line. All KFP juices and nectars are produced at the Kuryong Po facility, which in 1976 was running at capacity (340,000 cases of forty-eight cans of 250 grams, and 27,000 cases of the 1,360 gram size). The trade department of the company estimated sales growth at 30 percent over the next five years. Clearly, there was a need to add a new processing facility, which KFP intended to do by expanding the Taejon plant. The consultant reviewed a proposal that the company had received from a Swedish food processing equipment manufacturer and noted that the final cost would be closer to \$2.156 million than the quoted \$1 million. One reason for the higher cost was that the bid was a year old, but the principal reason was that all of the complementary equipment and auxiliary facilities had not been included. The offer included some equipment that seemed unnecessary given existing market demand for products and excluded another desirable production process. The consultant recommended that a process engineering analysis be made to stipulate the specifications and that several new quotas be solicited. He recommended that a high-speed line be purchased and temporarily installed at the Kuryong Po plant to increase the 1977 production, and that provisions be made to expand further the Taejon plant by 1980, when the new line might no longer be sufficient.

New Product Development. In his general market analysis, the consultant saw real potential for the introduction of new convenience foods in Korea. One half of the Korean population was now urban, as compared with one quarter in 1955, and the population, which had grown from twenty-one to thirty-five million, was expected to reach forty-one million by 1981. The population was young (more than half under twenty years of age), and most young wives worked. Urbanization would encourage the spread of supermarkets, which would offer the best display for the new products. So far, only 15-20 percent of families were using canned products with any frequency. Based on this analysis of the market, the volunteer encouraged the company to continue to expand its market testing of soup recipes, and defined the criteria and procedures to be followed.

Quality Control. Experiments were conducted under the volunteer's supervision to uncover certain quality problems. The volunteer drew up a Quality Control program which stressed weekly organoleptic sampling of production at each factory, a monthly check by the operations committee, a monthly quarantined product report, and a roundtable conference on quality control.

Results of Project. A year after the consultant's departure, some of his recommendations, in relation to the specific technical problem and general management, were clearly resolved; what is more important, each plant had developed a separate quality control section, with checks at every production step. The company was actively studying the sterilized pouch, used in Japan, as a substitute for cans for some of the products. Container savings had been realized by reducing the thickness of tin plate, but no new products have been introduced. The new juice line was purchased from Sweden, and delivery was expected in April 1978. The company was contracting with KIST to increase its process engineering capability.

Perhaps the most important result of the KFP project consultation was the strengthening of long-range planning capability. The visible shortcomings of the Busan project convinced the firm that its planning function was inadequate. Following the recommendation of the consultant, a Planning Department was established, reporting directly to the general manager. The plan to convert the Busan warehouse was abandoned, and the plan to extend the Pohang plant, where space was less of a problem, was on the drawing board.

INTERNAL FACTORS: HISTORICAL CHARACTERISTICS AND RESOURCES OF THE USER FIRM

The ability of a firm to receive and utilize technology depends upon the historical characteristics and resources of the firm as well as its general management practices. In examining the sixty-seven firms, special attention was paid to the following characteristics and resources:

- Technology acquisition history;
- Internal task technology;
- Organizational structure;
- Positive organizational myth;
- Charismatic leadership;
- Ownership;
- Strength of labor unions in the firm;
- Comprehension by management of the value of technology;
- Management resources;
- General manpower resources; and
- Financial resources.

The history of technology acquisition of a firm appears to be the most important, and potentially the most positive, historical characteristic in terms of the capability to absorb technology. Those firms that have experience in acquiring new technologies, and that in the process have learned to diagnose their technology needs and value technology, are better prepared to exploit opportunities for technology transfer. Other factors that can enhance the ability of a firm to receive technology are an organizational structure in which authority and channels of communication are clearly defined, a positive organizational myth, or charismatic leadership. Ownership of the firm may also be closely related to the outcome of technology transfer. There were in the sample of sixty-seven a sufficient number of firms in two ownership categories—government ownership and private companies with few owners—to allow conclusions to be drawn. An examination of the case studies shows that private companies controlled by only a few stockholders evidence change in the organization as a result of the project eight times as often as do government-owned companies.

The most important, and the most frequently cited, characteristics and resources that impede or discourage the acquisition and utilization of technology are (1) the lack of financial resources and (2) the limited availability of management and technical manpower to receive the technology.

The case of Bosnan Pharmaceutial Company (presented below) examines a charismatic leader in an organization with loyal and devoted employees and the importance of these factors in a successful transfer of technology.

Case Study: Bosnan Pharmaceutical Company (Seoul, Korea)

This case illustrates the importance of internal characteristics—strong history of technology acquisition, a charismatic leader, and a positive organizational myth—for a successful transfer of technology. Although the client firm already held a strong market position, it recognized a market opportunity for a new product and wanted to capitalize on its unique technological advantage to exploit this opportunity. The national government encouraged the company to experiment with the new product by protecting it from competition from foreign pharmaceutical companies.

The Bosnan Pharmaceutical Company was one of the earliest pharmaceutical distributors and manufacturers in Korea. When it began operations in 1946, it was involved in manufacturing yeast products, extracting ginseng for tea and extracts, and developing a digestive enzyme fermentation and manufacturing program suitable for the production of certain forms of antibiotics (especially penicillin). During the early stages, the company was supported by licenses from two Swiss companies and a British firm. By the end of the 1960s, the company was ready to launch a major expansion program with the aid of additional foreign licenses. Substantial debt was incurred in order to finance the new expansion plan. Unfortunately, the founder of the company, who had dominated all of its activities and whose spirit and managerial capabilities were the key to its early success, died at the time of the new plant construction.

In the next few years, two different managers tried to rescue the firm, but without success. By 1970, Bosnan was near bankruptcy and its sales and profitability had dropped dramatically. In 1971 a young enterpreneur acquired the firm. Bosnan's backers viewed the entrepreneur as the only hope for pulling the company through its financial difficulties, and the entrepreneur himself was confident that he could accomplish this because several years earlier he had taken over a chemical company that was in a situation similar to that of Bosnan and had built that company into a strong and viable organization. His reputation as a company saver helped him to gain the financial and other support required to acquire Bosnan.

As with the previous company, the young owner focused heavily on marketing through the development of new products. He also based his rebuilding operation on new manpower, recruiting aggressive young pharmacists and engineers and giving them the authority to experiment with new products, research techniques, and marketing strategies. In 1974, he acquired an electronic company that was involved in manufacturing capsulation equipment, calculators, radios, and medical electronics.

When the new president of Bosnan took over, he immediately began a search for the best antibiotic fermentation experts in Korea. Through personal contacts, he was able to recruit a key pharmacist in this area who had worked with a company under license with Spanish and other European firms for specific fermentation processes. The new manager and president decided that there was a significant market opportunity in the synthesis of penicillin. At the time Korea imported bulk penicillin, and very few foreign companies had seriously considered the possibility of manufacturing synthetic penicillin locally.

The top management of Bosnan took a substantial risk in allocating much of its limited resources to the new research and development system established to exploit the market opportunity for penicillin. The new research group and other groups in the company were actively involved in developing new management systems for the entire organization. Although production and sales techniques were being changed rapidly, it was felt that research and development was the key to gaining a competitive advantage that could effectively move Bosnan Pharmaceuticals forward. That breakthrough came in late 1973 when Bosnan became the first Korean company to develop a patentable process for synthesizing 6 APA penicillin. Eventually, the company succeeded in chemically synthesizing this product (an essential ingredient for ampicillin) from penicillin G, which gave the company high visibility and a significant market advantage within Korea. The government moved quickly to grant Bosnan import protection against other synthesized penicillins. The sales of the firm more than doubled in 1974. By 1975 sales were 600 percent of what they had been in 1973.

With the successful breakthrough in ampicillin production, Bosnan was able to expand rapidly its activities. It was already manufacturing and distributing the products of a U.S. company, and it began looking at other types of antibiotics that could be synthesized from its penicillin G development. Cloxacillin and a new type of combined antibiotic were developed in 1974. By 1975 the company began licensing its know-how outside of Korea. The 1973 breakthrough on penicillin G led to a new process for the synthesization of Amoxycillin, for which the company was awarded another patent. It was at this time that the company began to consider ways of upgrading its general fermentation production.

There are basically three stages in the production of antibiotics. The first stage deals with developing an effective culture that will generate the antibiotic ingredient. During the second stage, the culture is fermented and the antibiotic by-product is developed. In the third stage of recovery, the usable antibiotic material is filtered or otherwise removed from the fermented batch. Bosnan Chemicals realized that the potency (production efficiency) of its culture was inadequate. Through careful planning, company personnel were able to diagnose what type of help they needed to improve their overall production capacity at the culture/potency stage. Because the companies from which they were licensing other kinds of technologies could not provide this system-specific know-how, the company looked to other sources of assistance.

Since the reorganization of the firm in 1971, efforts had been made to gain competitive advantage through the development of new products. Whether hiring people away from other firms or utilizing licensing techniques to gain know-how, Bosnan had successfully created several market advantages. One principal area of production, enzyme manufacturing, had been idle for some time because the firm lacked the experience and specific know-how to make this area competitive. The Bosnan staff heard about IESC volunteers from the Korean Association for Small and Medium Business. After some investigation the firm was convinced that the consultant program could provide the system-specific know-how that it needed in the enzyme production area.

At the time of the consultation, the firm had over $6 million in sales (an annual growth rate of more than 100 percent) and employed about 250 people, of which some 25 were in management positions. Morale was high and management and technical personnel were enthusiastic about implementing more modern and efficient management techniques. The company had two full-time staff who did nothing but review manuals, publications, and other written materials available at the Korean Science and Technology Information Center. These two individuals regularly visited all of the major hospitals to examine newly developed products. It should be noted that the original breakthrough on the 6 APA penicillin synthesis had come from their review of a series of Dutch patents, which in turn triggered certain ideas among the pharmaceutical researchers of the firm.

Internal Firm Characteristics Affecting Technology Transfer. By 1975 the three company group was run almost directly by the new, young, charismatic president. The Bosnan Pharmaceutical Company was headed by the president and an executive director, who was primarily responsible for overall management and financial control of the company. The company had four functional departments reporting to the executive director; these were the sales department, general administration department, production, and international operations. The research and development and planning departments were the two major staff functions, and they reported directly to a general vice president of the company who in turn reported directly to the president. The two most important divisions of the company were research and development and the production department.

The production department had its own planning, control, and administration offices and was responsible for its own shipping, technical operation control, maintenance, library, and warehousing. Three product line groups were organized under the production department. One group focused almost exclusively on antibiotics; the second concentrated on synthesis, fermentation, and extraction; the third group produced ointments, liquids, and soft capsules. Quality control and packaging were separate departments outside production.

The research and development system was composed of a planning and control department that had two separate sections (technology development and product development), an office of market research, technical information, and medical information, a technology group related to manufacturing processing, a standardization group, and a clinical pharmacology group. The purpose of this system was to develop new technology, to develop new drugs, to provide an informational service for the company, to develop planning for export services, and also to provide planning and guidance to the medical electronics group.

At the insistence of the young president, the firm developed a three-year operating plan, but also a revolving one-year budget. These systematic and detailed plans covered all types of general environment, industrial environment, firm environment, and product contingency issues. Information on technology was acquired through manuals and textbooks, specialized equipment, and a continuous search for technical journals and patent information. For specific system assistance, they relied primarily on foreign consultants and foreign licensing affiliates.

Firm policies were formally stated. A company newspaper was issued each month to clarify policies as well as to communicate news and explain new product developments. Authority was clearly defined through charts and a continuous discussion of management practices and care was taken to keep lines of communication open. The main problem of the firm was not in its management system, but in its lack of manpower.

Because of the outside difficulty relating to shortage of manpower, the firm developed a variety of on-the-job training, formal in-company training, and outside of company special courses. Over forty specialized product training manuals have been developed within the firm itself to train and guide its staff.

The primary motivator in the company came from the direction and energy of its young president. All decisions were made by him and a few of the critical financial and pharmaceutical experts he had brought to the company. There were no other major generators of policy or decision criteria. The credible success of his first few years, both at Pacific Chemical and then at Bosnan, gave him an almost unquestioned authority and loyalty within the firm.

Though critical decision making was dominated by a small elite clique, there was a strong emphasis on participatory meetings and management staff programs throughout the company. In addition to this, there was a formal suggestion system functioning in the firm; individual staff were encouraged to criticize plans and suggest innovations and changes to the proposed programs developed by the senior management team.

External Environment Factors Affecting Technology Transfer. Competition among pharmaceutical companies has traditionally been very active. A firm can develop a new active ingredient that will immediately give it a commanding lead in a particular product area. As a result, there is a great deal of industrial espionage and corporate secrecy. In most cases the key to the success of a pharmaceutical company is a breakthrough in its own research and development efforts.

During the 1950s and 1960s, the Korean government had been anxious to stimulate the pharmaceutical industry. A number of laws were enacted to provide tax incentives and competitive protection for Korean firms that could develop products that were being imported. When Bosnan confirmed its breakthrough in the synthesis of ampicillin, it was able to secure from the government an import ban for all outside suppliers of 6 APA. The local competitors of Bosnan attempted to block the patenting of the Bosnan development. In fact, initial attempts by the company to protect the new technology met with the stiff opposition of the local competitors who tried to convince the government that Bosnan had not really made a breakthrough. After Bosnan had been granted the patent, the competitors tried to force the granting of licenses to them so they could enter this new production area.

The confusion that sometimes develops among different ministries, fueled by competing parties, also creates confusion in the industry. For some years, until 1976, many Korean firms felt that it was not worth the risk to invest in research and development since they could not be sure that they could protect the value of their breakthrough once it had been achieved. Regardless of the problem of government support and protection of the new technology, Bosnan was able to acquire a limited monopoly in its new product area.

With regard to manpower, the Bosnan Company felt that many of the graduates of the Korean universities were not well prepared in the rigorous disciplines of pharmaceutical research, product development, and manufacturing. As a result, the firm worked diligently to hire away experienced personnel from other companies. Skilled laborers were also hard to find, and for this reason the company instituted comprehensive internal training and manpower development programs.

Process of Consultation. The Bosnan Pharmaceutical Company had for some time been convinced of the importance of developing new products and improving its efficiency. In preparation for the consultation, it undertook an extensive analysis to identify precisely the kinds of expertise it needed. Firm personnel also organized themselves to work efficiently with the consultant. The combination of a well prepared and receptive firm and the right consultant resulted in a highly successful consultation.

The volunteer consultation was requested to improve the existing commercial production of a digestive enzyme known as polyase. Improvements in the production of this ingredient would not only lower the cost of the product but also reduce the cost of other products of which it was an ingredient. The volunteer consultant brought to Bosnan the system experience needed to improve overall production and to develop new fermentation products. The consultant was uniquely qualified to work in this area since he had spent almost twenty years with the Department of Agriculture as a microbiologist and later as an executive in one of its regional research bureaus. He had spent an additional twenty years as a director of a microbiological research program for a U.S. pharmaceutical company. Since his retirement in the early 1970s, he had been a professor of biology at Western Kentucky University. Not only did he have the system-specific experience relating to fermentation of the enzyme, but he also had a background of teaching and consulting.

To improve effectively the cultivation of the strain for enzyme production, the consultant had to structure a program with special emphasis on diagnosis. The three-month work program covered four basic stages: (1) analysis of all steps of the production process (2) development of a plan for improving production areas related to the major losses of enzyme recovery; (3) development of new production solutions in a laboratory test environment; and (4) development of a full-scale pilot test for the entire production system.

The fermentation of the enzyme went through a number of stages. The sequential mutation and strain multiplication could be effected at any one of a number of stages relating to the development of the culture, fermentation, or final recovery of the enzyme. The consultant began his work by carefully analyzing each step of the production process to uncover problems. He identified a number of production techniques that had to be changed, including fermentation temperature, refining procedures, and crushing of raw materials used in fermentation. His suggestions for eliminating certain refining steps and modifying the fermentation activity raised the level of recovery efficiency from 25 to 75 percent.

The desire of the company to take advantage of the know-how of the volunteer was demonstrated by the speed with which the staff provided him with new machinery and by their implementation of his suggestions during the project. The company was oriented toward establishing market opportunities, and they knew the volunteer might be able to help them reestablish their company as an active producer of digestive enzymes. In short, the consultant had the complete backing of the company and an appropriately trained staff to receive and utilize his ideas.

The main strategy of the consultant can be characterized as catalyzing activity and prescribing new solutions. It is important to note that the consultant used a small team of local engineers/pharmacists to work with him through a detailed diagnosis of the entire production system. From this diagnosis, he isolated the sources of problems and prescribed a variety of solutions. In suggesting solutions, he conducted extensive training on how to implement the new procedures. Although the project was primarily directed at instituting new controlling mechanisms in production, the consultant also helped the firm to establish a variety of techniques and systems that could be used in the ongoing research and development activity. For example, he worked with several groups in setting goals for research and development, designing daily work flow control documents, and other research and development management programs. He also helped to orient the firm toward outside sources of information and was especially useful in encouraging its collaboration with local universities.

Results of Project. To support his overall change activity, the consultant used an analytic approach for diagnosis and information feedback to the research and development group working with him. He prepared working memoranda and a final report on different product techniques and product developments including:

- Production of fruitose or fruitose syrups;
- Procedures for the measurement of pectin hydrologizing enzymes;
- Dilutants suitable for use with enzymes;
- Digestive aids;
- Recommendations for operating conditions for the enzyme plant;
- Evaluation of mutation programs;
- Guidelines on product recovery steps of precipitation and washing; and
- Recovery of special cultures of fungi and bacteria.

These memoranda are still being used by Bosnan, as are a variety of management aids provided by the consultant. The most dramatic result of the project, however, is the viability and cost effectiveness of polyase digestive enzymes, a product that the firm is now actively producing. Bosnan introduced this product in 1977 to the market directly and has already been using this product as an intermediate in a number of its other production activities. The potential dollar value profit of the direct sale of polyase could be as much as $250,000 in 1978 alone.

In planning for the research and development department, the consultant also left behind an analysis of work that could be done in the future to further improve fermentation production. He identified a number of problems resulting from contamination of cultures with other fungi, improper procedures for mixing alcohol and extracts in the precipitation tank, inadequate control of temperature for different extracts, and improper timing and coordination of recovery processes, and suggested techniques for solving these problems.

He also developed more advanced research and development goals relating to the development of new chemical enzyme activators, improvements on the culture inoculum, and programs for analyzing the maximum time for incubation.

In summary, the consultant not only provided the know-how for more efficient production, but also provided guidelines for improving the overall research and development capacity. His identification of other problems and potential research and development opportunities could lead to other significant breakthroughs for the company. What is important is that this company was led by an able and enthusiastic president who understood the value of foreign assistance and technology and was already market and research oriented. Personnel were thus more willing to implement vigorously the suggestions and ideas of the consultant. (Bosnan is considering bringing in another volunteer consultant to further develop its research and development capability.) The success of this project depended to a degree on accurate diagnosis of a specific production and product requirement; this facilitated locating a consultant who had the specific capabilities to address the problems identified. The company was able to move more quickly to implement the recommendations of the consultant than would have been possible in the absence of accurate diagnosis by the client.

EXTERNAL FACTORS: EXTERNAL LOCAL ENVIRONMENT

Each firm exists within an environment whose characteristics affect positively or negatively the operations of the firm. Support for the firm from the technological environment, in the form of readily accessible laboratories and suppliers, has a significant impact on how managers look beyond their firms for solutions to problems. Government policies and the general confusion resulting from frequent and unpredictable policy changes clearly affect the ability of the firm to plan for long-term development and, hence, to acquire technology.

In examining the sixty-seven firms, the following external environment factors were examined:

- Market,
- External competition,
- Environmental turbulence,
- Governmental action,
- Source of supply,
- Local technological infrastructure,
- Cost of technology, and
- Appropriateness of technology.

The most positive of the external factors in effecting change in managerial practices is the market and the demand it creates for products, which in turn encourages the firm to make certain technological improvements. External competition is also a positive factor since it can force the firm to change and adapt to new developments within its particular industrial sector. In some cases, policy has inhibited change in firm behavior. However, policy (particularly if it does not vacillate or change frequently) is generally seen as a positive factor since it encourages the firm to adapt to external factors.

The principal external constraint to technology acquisition and utilization is the weakness of the local technological infrastructure. In most cases, the local firm has very few sources to draw from for the skilled managers it needs to exploit market opportunities or to improve its competitive position. Frequent changes in governmental policies tend to create an unstable and unpredictable environment in which the firm is discouraged from taking new initatives in expanding or improving its operations. To a lesser extent, the cost and the appropriateness of technology also act as potential negative constraints on technology transfer.

Environmental turbulence—that is, how rapidly the environment is changing and how well the firm understands these changes—is a special issue that is examined more closely in Appendix I. At this point, its importance need only be highlighted. Those countries in which the government frequently changes its policies and commercial laws are viewed as unstable or turbulent environments. Another indicator of general environmental turbulence is the degree of difficulty a firm has in obtaining information about governmental policy.

The firm may also be affected by the instability of its particular industry group. For example, those industries involved in production and marketing of consumer goods may be more severely affected by shifts in market demand than those in more stable industry groups such as equipment manufacturers.

Diagnosis and goal setting, both of which are critical to effective technology transfer, require some degree of stability in the external environment. The perception by the firm of an unpredictable or hostile environment must be offset by positive external factors if the firm is to seek and acquire new technologies.

All but two of the five countries in which field research was conducted are characterized by turbulence and rapid change. Peru is clearly the most turbulent country, with Brazil and Tanzania ranked as second and third most turbulent. Kenya is a relatively calm environment for the user of technology, and Korea is the most calm of our sample. It is interesting to note that this classification of the five countries supports the findings concerning the relationships of the general environment to the ability of the user firm to plan and utilize technology. For example, a high ratio of transfers that were not successful are found in Peru and Brazil while in Korea and Kenya, more often than not, projects led to change in the behavior of the firm.

The impact of the market on the receipt and use of technology by a firm is readily discernible in the case of BoaStruct, which is presented in the following pages.

Case Study: BoaStruct (Rio de Janeiro, Brazil)

This case demonstrates the ease with which technology can be transferred when there is an effective receiver. The client firm, which had a history of technological development and importation, was well prepared to receive the engineering skills of the volunteer and to use these skills to establish a new product and service activity. The characteristics of the industry and the local environment led to rapid external diffusion of the technology to government agencies and competitors, resulting primarily from the demonstration effect of the capabilities and characteristics of the new product. The specific know-how transferred did not lead to a dramatic change in the management activities of the user firm, which was already quite sophisticated in its overall management orientation and procedures. As in many cases in the Third World, immigration played an important part in the establishment of BoaStruct. The father of the present chief executive officer came to Brazil after the Second World War and started a small business that supplied suspended scaffolding to the construction industry, acting as the agent for an American firm. His son, following his education in the United States and service in the American military, joined the business and in 1964 bought out the interest of the foreign joint venture partner. At this time the total sales of the firm were only $75,000. Three years later, sales had reached $180,000 and the firm introduced a type of tubular scaffolding to replace its suspended scaffolding technology. The rapidly expanding Brazilian economy allowed for more growth, and in 1968 the firm began initial manufacturing of panels and frames for its scaffolding. Local tubing was purchased and then cut and finished in the plant. The growth potential of the firm encouraged the owner to travel abroad in search of new ideas and techniques for his business. Between 1969 and 1972, a number of technological changes were introduced, especially in the area of stronger and more efficient joints for the framing. In 1974, the firm reached a sales level of $1.5 million, which grew to over $4 million by 1976. The BoaStruct work force increased rapidly from 1974 to 1977, at a rate of over 20 percent per year. By 1977, the firm employed 330 permanent employees.

The technological history of the firm seemed to follow the growth in sales and demand. The principal start had come from the American firm, whom the family represented until 1964. Suspended scaffolding technology provided the base for growth and change. The senior management of BoaStruct later traveled extensively to Canada, the United States, and Puerto Rico to visit plants and look for new systems. Many foreign firms opened their doors to them since BoaStruct was small and the companies were interested in the possibility of future joint ventures with BoaStruct. A Canadian firm permitted BoaStruct management to film one of its plants. The Brazilians were thus able to bring back an entire plant on film, which was used to improve manufacturing and service capabilities.

In 1974-1975, BoaStruct was approached by two American firms and a Japanese company, which were interested in establishing joint ventures. The Brazilian owner decided against this approach since he felt that the firm could get the technology from other sources. Although the company needed additional capital for growth, the management decided to wait until it was absolutely necessary before they gave up any ownership or control. By 1976, BoaStruct was quite large and foreign firms were not as willing as they had been earlier to share information and concepts with the company. Foreign enthusiasm for participation also diminished as the Brazilian government began imposing restrictions on foreign capital and technology.

To expand into new product areas, the firm had to find a source of the technology suited to this specific process. The sales department and senior management knew their steel frame system was cheaper and more effective than the tube and coupler systems of their competitor, which had come from Europe. The staff believed that it could use this technological base to move into other areas, such as shoring and elevator towers. BoaStruct found a Swedish firm to supply motors and license other know-how for the elevator towers, but the Brazilian government refused the contract; the government argued that the technology already existed in Brazil. BoaStruct disagreed with this assessment and did not trust the local system identified by the government. Knowing the difficulty of changing the views of the government engineers, who did not understand the complications of the industry, BoaStruct sought other technology alternatives. The firm learned that IESC might supply a trained specialist. Using this expertise, the company could move into the shoring area without licensing techniques or equipment and without forming a joint venture with a foreign firm, thereby avoiding government bottlenecks. After discussions with IESC, an expert with over twenty years experience in steel frame shoring was located and contracted for consultation.

Historically, the firm has not had a problem finding technology when it needed it. Rather, its problem has been that of finding management personnel and technicians who can absorb and manage technology. This problem has persisted even though the firm has spent significant resources in supporting the education of many engineers and staff. The problem seems to hinge on developing firm-specific and system-specific expertise for the business. The president of BoaStruct now believes that as the firm continues to grow, he may have to form a joint venture with a foreign firm to stabilize his access to technology, as well as to generate new capital, which is the other major problem inhibiting growth of the firm.

Internal Firm Characteristics Affecting Technology Transfer. The principal activity of the firm centered around the sales and servicing of scaffolding and shoring systems. Though it manufactured most of its components, the firm was service oriented since the critical aspect of the sales related to designing and setting up the equipment. The company was organized by function, with two of four divisions focusing on sales and servicing. The other two divisions dealt with manufacturing/maintenance and finance. A total of eight other staff groups dealt with safety, technical assistance, accounting, costs, and auditing. Since the firm was trying to develop a more effective staff structure, it had not yet rationalized its various staff functions into an effective staff group that could deal with planning and control of the overall firm.

The formal structure of the firm was hierarchical, though there appeared to be a significant amount of group participation in most decision making. The president was anxious to delegate authority when possible in order to develop managerial capability. The structure of the firm was identified in formal charts; job descriptions existed for key personnel.

Management staff comprised about twenty-five individuals who participated directly in decision making. Great emphasis was given to recruiting and developing management. The company directly financed the formal training of a number of engineers, without any mutual commitment regarding employment after graduation. Salaries were high and the staff evidenced an enthusiasm related to the ability to participate in decision making. The president stated that "to motivate my managers, I have to ensure them an opportunity to contribute and also respond to their ideas—they have to have an echo inside the firm." A variety of formal training methods were used internally to continue the development of management. Outside courses were also encouraged and paid for by the company.

The structure of the organization evidenced the strong commitment of senior management to technology acquisition. All engineers who were hired were required to speak English so that they could utilize journals, foreign visits, and consultants. Two local consultants supplied general administrative support on an ongoing basis and a formal consulting board had been instituted. This outside committee met regularly to evaluate the technological condition of the firm. To ensure that management kept abreast of general industry developments available to Brazil, a patent lawyer was also kept on retainer to review all new patents and management agreements filed with the government and inform the management of any new developments. The president also worked closely with a number of foreign associations to get new concepts. Local seminars provided a variety of general management programs. The programs using foreign professors or technicians seemed to be the most useful, although they were quite expensive. Equipment sales to the company did not represent any significant source of new ideas and concepts.

Obstacles to technology transfer related to the evaluation and selection of alternatives and locating management that could use the know-how once imported into the firm. The extensive variety of local consultants presented problems since the firm could not distinguish the good from the bad and could seldom get consultants relating to its specific operating problems.

"For our firm, planning is the biggest problem," concluded the president. At the time of the consulting project, BoaStruct had hired several senior financial staff to develop a planning and staff leadership team. It was planned that the different staff sections reporting to the president would ultimately come under a chief of planning and financial controls. Planning was limited to annual operating budgets which were reviewed periodically. These plans did not formally deal with contingencies of alternate government or industry developments. Technology planning was minimal, though the firm had consciously developed a variety of information sources.

Standards and procedures were developed by top management and relevant staff. The president acted more directly in the area of marketing and administration than in production. Cost controls existed for a number of production activities, but not for the entire company.

Though the company was formally centralized, delegation of authority was encouraged. From the president to the laborer, decisions flowed through five to seven levels of authority. The president participated in all staff and senior management meetings. His participation was especially active in the sales and administrative areas. When a staff meeting was held in the absence of the president, tapes were made so that he could review them. The president apparently felt that his response to questions and discussions was necessary to maintain interaction. To encourage suggestions and creative thinking, awards were given for the best recommendations from all levels. The managerial staff felt they could disagree and freely suggest alternative courses of action, though it was unclear what authority they had to carry through ideas or programs once developed.

All policies were set by the president and relevant staff. No formal mechanism existed to communicate these policies, but it was felt that the informal transmission by staff and informal meetings was an effective means for passing on policy developments.

External Environment Factors Affecting Technology Transfer. Since the mid 1960s, Brazil has been a fertile environment for business. The construction business has grown dramatically. An atmosphere of relative laissez-faire prevailed until the mid 1970s when the government began restricting imports and constraining relations of local firms with foreign sources of capital and technology. Internally, the government had not impeded private industry, and in terms of developing managerial and technical capabilities, a number of incentive systems and assistance programs were established. Controls were periodically introduced to control inflation and balance of payments problems, but in general the firms were allowed to grow as fast as possible. In particular, the scaffolding industry had good contact and communication with the government.

In terms of technology, the local environment provided adequately trained personnel at the university level. Locating experienced personnel was, however, very difficult. The industry itself was too young to have a wide scope of local expertise, and although the environment provided consultants in a variety of general management areas, specific technical assistance was scarce.

Contact with local universities and the government existed, but without much benefit to the local firms. For example, although BoaStruct personnel carried out material tests at the local universities, they supplied the specifications and guidance. The government allowed private firms to participate in its norms and standards committees but had not generated adequate industry guidelines. The IESC project left behind standards and norms, particularly in the area of safety, which were passed on to the government to aid in developing more effective national safety guidelines. In fact, as a result of the consultation, BoaStruct formed a group of local competitors to rationalize the European, Japanese, and American approaches and propose a common set of standards and norms for the government.

Regarding substantive issues in technology, the local associations were not effective. Their major focus was on political interaction with the government. The contact with foreign associations and industry groups apparently provided important information and contacts. The president of BoaStruct belonged to the American Scaffolding Association and another California industry group. For some reason, these contacts are discouraged by the government; payments for foreign membership are not allowed as a business tax deduction. Even with this obstacle, local firms apparently join a variety of European and American groups to gain technology inputs and industry information.

In summary, cooperation existed between the firms and the public sector but it had not grown into a major source of support for the firms. While the concern of the government over foreign technology had cut off certain opportunities for growth of local firms, in other areas related to encouraging development of local firms, tax incentives were given for in-house training (a double tax write-off could be received for a program approved by the government) and government agencies were available for administrative counseling. A variety of local consultants was available, but it was difficult to determine which had the appropriate capabilities to support the firms. Though technology support was growing, there was a lack of knowledge on the part of both the government and the firm concerning the selection and application of technology.

Process of Consultation. While the local environment provided a number of sources for general managerial technology, locating system-specific experience was quite difficult. Scarcity of managerial talent also made it difficult to create an effective group for receiving technology. Without going directly to foreign firms for licensing and joint venture arrangements, the local firm had few options for locating appropriate specific assistance. The location of an IESC expert with the exact industry experience needed by BoaStruct enabled the firm to enter a new product area and establish some general planning systems and goals for the company.

The request for assistance related to establishing a new product area (steel frame shoring) to complement the existing technology base in steel frame scaffolding. The primary goal was to help adapt the firm's equipment and prepare the firm for marketing to the construction industry. This meant assisting in training and planning activities, as well as creating standards and techniques that could be used to change the practices of the customers. The principal users of the know-how were the two engineers and senior management involved. The indirect users were the customers and government agencies who would be approached after the initial company development was complete.

The strategy for change was to catalyze the existing management into a new team and prescribe specific plans and standards for the new activity. The consultant spent about four weeks analyzing the firm and the local environment. The rest of his three-month visit was spent devising program specifics and training the staff in the modified systems. Specifically, the target program called for the following:

- Previewing company equipment, facilities, and staff capabilities in light of existing United States methods;
- Visiting the local construction sites to observe and evaluate shoring activities, local materials, and other construction practices;
- Designing new (or altering existing) systems for scaffolding for application to shoring requirements;
- Developing shoring sales and methods manual;
- Training staff in planning and maintenance of shoring installations;
- Developing a long-term program for future entry into more sophisticated types of shoring, such as horizontal shoring, rolling shores, angle shoring methods, heavy duty bridge type shoring units, modular steel shoring forms;
- Creating safety standards and guidelines suitable to the entire industry.

Information delivery as part of the change strategy relied primarily on informal meetings and group discussions of problems and situations identified by the consultant. Though no formal feedback was used, the consultant did informally analyze gaps between existing systems and desired situations and presented these analyses to management in order to clarify program steps. The focus of change was on middle management/engineers, but there was a strong impact on the president and some spillover to the manufacturing staff. For example, the consultant observed a number of fabrication practices that needed improvement, and spent time on the factory floor demonstrating new drilling techniques or machine handling procedures. Though some production know-how was transferred in these situations, the heart of the program lay in establishing the specific plans and programs for the new product area and the related controls to guide the service and manufacturing activities.

Results of Program. The results of the program were very dramatic. Within two weeks of the completion of the project, BoaStruct began its first job in shoring, and the marketing department was advertising broadly the characteristics of the new product service. The reduction in costs and increases in safety and efficiency were quite obvious to the market. Within six months of the entry of BoaStruct into the market with the panel shoring, three local competitors were bringing in Japanese and French consultants to install their own type of steel frame shoring.

The work of the consultant with safety standards was also rapidly diffused to the external environment. BoaStruct had earlier played an important part in establishing initial safety norms for scaffolding. As a result of the consulting project, BoaStruct established a new program with its local competitors to rationalize the new shoring technologies so that they could propose more specific standards to the government.

Besides producing the necessary system specific know-how for the new service, the consultant established long-term planning goals for a number of related services. This will enable BoaStruct to initiate a systematic expansion and introduction of new products based on its growing technological and managerial capacities.

In summary, the impact of the program can be viewed as follows:

- Developed a plan and guided redesign of existing equipment and design of new accessory items for shoring work including samples and development of initial production stock;
- Trained engineers and sales personnel in methods for planning specific shoring jobs and how to develop shoring quotations and controls;
- Aided in hiring new sales engineer and his initial preparation;
- Developed example engineering charts, formulas, procedures, and safety data;
- Established manual for steel frame shoring activities including sales guides, shoring tower tests, load factor guides, guides for developing engineering shoring layouts, guides for lateral bracing, comparative cost guides, safety rules and guides for future shoring development programs;
- Formulated planning guide for new service area analysis.

BoaStruct is now using the shoring manual extensively in daily operations and preparing new technicians. The emerging company planning effort will use the suggestions of the consultant on goals and possible new product areas as aspects of their long-term strategy development. The savings due to one improvement (a new form of connecting pin) gained from the volunteer will alone reduce costs by over $50,000 in the first year of its use.

Although the overall management capability of the firm was not changed, the development of the new service activity will benefit both the firm and the country by increasing safety and efficiency, while reducing costs. The culture of the organization was already quite open to change before the consultant arrived and personnel were anxious to develop new management systems.

The forces working against change are linked to the problem of available manpower. BoaStruct had hoped to have a new civil engineer to work with the consultant, but could not locate one until the last weeks of the project. This significantly reduced the potential impact of the project, but the training of the other engineers was passed on to the new man after the consultant departed. Due to the effectiveness of the client firm in absorbing the technology, the firm will not need follow-up assistance to aid in implementing the specific project recommendations.

The client firm was well prepared to work with outside sources of technology and knew what it needed. The specific experience of the volunteer covered a variety of system-specific knowledge in the scaffolding and shoring industry, so the match for transfer was very appropriate. Because of the specificity of the know-how and its target use, little change was noted in the rest of the firm, though the volunteer did provide some inputs into establishing a long-term growth plan. Adaptation of the volunteer's knowledge was not difficult since his experience was so broad. Technology acquisition by BoaStruct was inhibited by the government in some areas, but was supported in others, such as in-house training and cooperation on national standards development.

EXTERNAL FACTORS: TYPE OF TECHNOLOGY

Technology is a complex concept that connotes many things to many people, and establishing a single typology that addresses all of its characteristics is difficult. One approach to viewing technology is to define its general characteristics and to describe the general concept. A second approach to viewing technology, which is useful in illustrating the concept, is to describe the context in which technology is applied in the solution of specific problems. In this study, the context was the individual consulting project. An examination was made of both the general or conceptual characteristics of technology and of the contextual or situational characteristics—the characteristics of the consulting project—that led to a certain type of technology being applied.

The following summarizes the approaches taken in this study to viewing technology:

Conceptual Characteristics
- Technology as a type of knowledge;
- Technology as a management process component.

Contextual Characteristics
- Technology in terms of the target of consultation;
- Technology in terms of the events in the firm or in the external environment triggering the consultation;
- Technology in terms of other attributes relating to the results of the consultation.

Examining technology from these different perspectives demonstrates again the need for concentrating on both general and specific technologies in a consultation project, and the importance of planning, in developing user capabilities to acquire and use technology.

Conceptual Characteristics of Technology

Technology as a Type of Knowledge. Technology may be defined as the knowledge or know-how necessary for the productive functioning of an organization. This can be broken out into five basic categories:

1. *General knowledge:* that which is publicly available to the society (for example, through books and universities);
2. *Industry-specific knowledge:* that which is necessary to produce a product or manage a process which is generally known within the industry (for example, how to produce and market pharmaceutical products);
3. *System-specific knowledge:* that which is necessary for the production of a specific product (for example, antibiotics);
4. *Firm-specific knowledge:* that which is necessary to produce a product or manage a process that is owned by or contained within a specific firm;
5. *Ongoing problem-solving capability:* Know-how that results from experience and that is necessary to solve production process problems. This capability is linked to general knowledge but combines elements of the other three types of more specific knowledge.

It might be assumed that in the interest of gaining competitive advantage, most technology acquiring firms would want to acquire system-specific or firm-specific knowledge. The case studies of developing country firms show that while the client most often requests this kind of technology, the volunteer working on the project places greater emphasis on the more general kinds of knowledge. Seventy-nine percent of the 472 volunteers responding to the questionnaire stated that their projects dealt primarily with a transfer of industry-specific knowledge and on-going problem-solving capability despite the fact that they were requested to provide system- or firm-specific knowledge.

Table 3-9. Project Characterized by Knowledge Categories (N = 472)

	Consulting Project Dealt With	
	Number of Projects[a]	*Percent of 472*[a]
General business knowledge	278	58
Industry specific	372	79
System specific	222	47
Firm specific	220	46
Ongoing problem-solving capability	373	79

[a]The totals of these columns exceed 472 due to multiple classification of projects.

Technology as a Management Process Component. The second of the conceptual characterizations of technology relates to the process of management activities. The great majority of projects are initiated by requests for organizing or controlling activities to help deal with specific problems or to exploit new market opportunities. The case studies show that while the volunteer enters the project to work on these specific issues, he soon moves the client toward planning and leading issues that must be addressed if his recommendations for new techniques are to be implemented and sustained by the firm. An analysis of the 472 volunteer questionnaires reveals that more projects addressed planning and leading (activating) than organizing and controlling.

Table 3-10. Process of Management Technology (N = 472)

	Number of Projects[a]	*Percent of 472*[a]
Organizing/structure	169	36
Planning	237	50
Controlling	188	39
Activating/leading	201	42

[a]The totals of these columns exceed 472 projects and 100 percent due to multiple classification of projects. For example, general business knowledge and system-specific knowledge may both be addressed in a single project.

Contextual Characteristics of Technology

Technology in Terms of Target of Consultation. From the project requests and the understanding of the short-term profit motivation of most developing country firms, it might seem that the technology transfer would more likely result from projects that focus on the introduction of new products, services, or production processes. However, responses to the 472 volunteer questionnaires indicate that production processes, values of the firm,[3] and motivation patterns are cited more frequently as project targets.

Successful transfers are found to occur more frequently in projects that deal with motivation, production processes, and values than in those that are concerned with new products or services or general organizational structure.

Table 3-11. Project Target (N = 472)

	Number of Projects[a]	*Percent of 472*[a]
Product/service	173	36
Production process	216	46
Organizational structure	176	37
Values/policies	208	44
Motivation patterns	203	43

[a]The totals of these columns exceed 472 due to multiple classification of projects.

Technology in Terms of the Events in the Firm or in the External Environment Triggering the Consultation. The events that trigger the need for a change in a firm's behavior are another dimension of the contextual characteristics of technology. The internal situation of the local firm that triggers search for and acquisition of technology might include a decision to introduce new products or the need to respond to a crisis situation. When a firm decides to take advantage of new market opportunities, to capitalize on internal capability unique in the local environment, or to respond to a government incentive, it is responding to external environmental triggers.

Analysis shows that almost half of the sixty-seven projects in the smaller sample resulted in a firm's request for assistance in modifying existing management or production procedures to improve efficiency. With respect to technology transfer and the consequent change in the firm's behavior, these projects result in a success/failure ratio of six to one. Projects resulting from crisis situations resulted in a successful transfer of technology in ten of eleven instances.

Table 3-12. Client Need that Triggered Project (N = 66)

	Number of Projects	*Percent of 66*
Crisis situation	11	17
New product	20	30
Modify existing activity	32	49
Defensive	2	3
Other	1	1

The analysis of the external environmental conditions that triggered the project indicates that firms seek new technology most frequently in response to a recognized market opportunity. Government incentives and a firm's desire to capitalize on whatever unique capabilities they might have are equally important in the search for technologies.

Table 3-13. Environmental Characteristics that Triggered Project (N = 67)

	Number of Projects[a]	*Percent of 67*
Market	34	50
Unique Technological Capability	15	22
Government Incentive	16	24
Government Disincentive	1	1
Other	20	29

[a]These totals exceed sixty-seven projects and 100 percent because of multiple classification of projects.

Technology in Terms of Other Attributes Relating to the Results of the Consultation. A third dimension of the contextual characteristics of technology is the other attributes relating to results of the consulting project. An analysis of the 472 volunteer questionnaires indicates that successful transfer of technology occurs most frequently as a result of projects where the techniques introduced are compatible with existing production or management procedures, where specific bottlenecks are removed, or where the firm saves time or money through use of the techniques. The attributes least related to successful transfer of technology and change in the firm's behavior are the element of risk in using the techniques and the ease with which the firm can return to its old systems and procedures. This illustrates again that, while firms may understand the value of new techniques for short-term improvements, they may be less convinced of the value of technologies that require a strong commitment to change in the management practices. (See Table 3-14.)

Conclusions. Several conclusions emerge from this examination of various types of technology. In the category of technology as knowledge, it may be seen that firms seek explicit technologies (system-specific, firm-specific), although the more general management technologies are more related to technology transfer since they contribute to an internal ability to identify problems and plan appropriate modifications. The controlling and organizing technologies may lead to easier solutions to the problem, but those projects involving planning and leading processes generate more permanent change in the receiving firm. In addition, it may be seen that those technology transfers that combine organization structure changes, personnel motivation, and value changes are much more useful than those that deal only with the specific modifications of systems such as production. The client needs most associated with technology transfer are those linked to external market needs or to modifications in existing systems for improved efficiency.

The consultant to Villa Clothing Company combined general management techniques with system-specific production skills and motivated personnel to change their traditional sewing methods. The following pages describe the consulting project and the resulting changes in the organization.

Case Study: Villa Clothing Company (Lima, Peru)

In this case study, firm management was concerned about production technologies yet benefited dramatically from the changes brought about by the volunteer's recommendations, the general values, and the motivation of personnel. The project, which was initiated in response to an internal crisis and an external market opportunity, resulted in changes in long-term planning as well as the solution of specific production problems.

Villa Clothing Company began in 1963 as a small shop with a few sewing machines. By 1977, it had a combination of plants utilizing over 400 workers. Sales are now over $1 million in the main plant alone. The company accounts for some 8 percent of the total casual clothing market in Peru, and in some product lines controls as much as 75 percent of the market. Growth of the firm was strong until the onset of the most recent economic crisis in Peru in 1977.

Table 3-14. Other Attributes Relating to the Results of the Consultation by Transfers of Technology (N = 472)

Attributes	*Number of Cases Successful*	*Number of Cases Unsuccessful*	*Number of Projects*	*Percent of 472*
Requires continuing expenditures	138	30	168	35
Clear-cut pay-off	158	30	188	39
Outsiders felt project necessary	145	21	166	35
Saves time	187	34	221	46
Saves money	185	30	215	45
Reduces physical discomfort	75	14	89	18
Bottlenecks removed	172	28	200	42
Use of ideas was risky	10	3	13	2
Ideas are compatible	176	29	205	43
Requires commitment	100	21	121	25
Firm can return to old systems	74	32	106	22
Opened firm to new innovations	120	27	147	31
Many personnel must participate	145	28	173	36

The growth of Villa, since its establishment fourteen years ago, is the result of the effort of one entrepreneur. In the late 1940s, the founder returned from the United States, where he had worked for several years at odd jobs in the clothing business, and set up a small shop to produce garments locally. During that initial period, he received important technology assistance from a consultant provided by a U.S. government program. This consultant helped the owner to establish manufacturing procedures that were used by the firm for a number of years and instilled in the owner a belief and interest in management technologies. As the operation grew, the owner searched for new technologies, primarily through his contact with equipment salesmen and visits to firms in other countries. In the late 1960s, he contemplated the use of a volunteer consultant to repeat the experience that had been so fruitful in his first years. The change in the Peruvian government in 1968, and the resulting uncertainty about the business environment, delayed this decision until 1976. Although the external environment was becoming increasingly turbulent, making the importation of technology more difficult, the owner sensed that he had many problems that could be dealt with by bringing in a consultant. Also, he had lost his plant manager in 1976 and thought that such a consultation would be useful in training the new plant manager in addition to helping reorganize production.

The client had other specific reasons for seeking assistance in 1976. While the firm had been known as a manufacturer of quality goods for some time and had a significant market share in this area, the management acknowledged that the reputation of the company was slipping and that there were problems with the consistency of quality. The firm had just acquired a new senior manager from a competitor whose operations had suffered from a loss of foreign technology. The manager was very enthusiastic about regaining contact with outside technology sources, and in this sense reinforced the positive orientation of the owner toward the consultation.

Internal Firm Characteristics Affecting Technology Transfer. The structure of the firm was hierarchical, with the owner as the dominant individual. Organization charts did not exist, nor were responsibilities formalized in job descriptions. The firm was organized by function into the areas of accounting, production, and sales. Workers and staff were involved in some decision making, but not in major policy or planning activities. A formal program for staff had not been developed though it seemed to be a concern of the senior management. Suggestions and complaints were handled in an ad hoc manner. The lower level workers were viewed with slight suspicion because of the growing union power in the national environment and the creation of the industrial community system. Decisions were made and policies formed by the owner, with ad hoc participation of key staff.

Growth and modernization seemed quite important to the leadership of the firm. The importation of new ideas and technology was stated as a key factor for business success. Visits to foreign firms and foreign industry magazines were major sources of information. Local business associations within the industry provided little aid in management issues but were looked to for discussion of government action and long-term market possibilities. Competitor contact was very limited and ideas were transmitted only through the success of a new product or systems that became highly visible.

Systematic controls were in place only in production but were not working efficiently. Production control lacked schedules and the forms needed revision. There was no control of inventory in process. Plant layout was a problem and reflected the ad hoc growth of the plant. Methods engineering for developing accurate costs, estimates, or controls did not exist.

No formal planning existed except in the form of rough annual budgets. Reviews were sporadic and most strategic thinking was informal. The scope of the budget was limited to sales and production and normally covered a period of one year.

External Environment Factors Affecting Technology Transfer. The external environment placed a number of constraints on the technology search of the client, primarily as a result of governmental policies regarding labor and economic development. An early example of governmental impact came in the area of ownership and control. The Industrial Community Law required all manufacturing firms to set aside part of their after tax profit for the labor group of the plant to finance their purchase of up to 51 percent of the stock. Since tax incentives were also available for reinvestment of earnings, the management accelerated their capital improvements to avoid taxation and slow down the turnover of the company to the workers. In the early 1970s, as inflation and dwindling foreign reserves became a problem, the government began restricting capital improvements through importation of technology.

The Labor Stability Act also created managerial problems since it fixes salaries at all levels and makes it almost impossible to fire any employee. While the firm used to take on extra workers at peak seasons, it now must be quite judicious in hiring, due to the permanence of any employment decision. The result of these governmental fluctuations is a sense of confusion on the part of management and a resistance to hiring or promoting. The stability laws inadvertently discourage the use of incentives, though many workers are still paid by the piece.

The primary source of management comes from the general training programs of the universities. Industry- and firm-specific knowledge is picked up through experience and contact with the industry developments with special training outside the country. Equipment salesmen provide an ongoing contact for ideas as do the publications from Europe, United States, and Japan on textile manufacturing.

Ideas on new techniques are transmitted primarily by success of products, new designs, or marketing systems. There is little direct information exchange within the industry, and industry associations act more as lobbying groups than as sources of industry- or system-specific technology.

Process of Consultation. The client accurately identified his needs by stating that "we feel and see a great need to update ourselves in production, especially in the areas of cutting and quality control." Since the fundamental stages of the activity of the firm are selling, cutting, sewing, pressing, and shipping, the client was requesting a control program for the overall production effort. The client also recognized the problem of leadership and requested a consultant who could "inspire our young engineers to be leaders and motivators, as well as technicians."

At the time of the consultation, the firm had an adequate design effort linked to its sales effort and a basic production system that could manufacture the goods in the cut, sew, press, and ship stages. Sales and marketing seemed strong, but quality was lacking in production. The initial diagnosis by the volunteer revealed no systems for quality control, raw material inventory, inventory in process, or methods improvement. Plant layout efficiency and organizational communication seemed inadequate as well. The project went beyond merely prescribing systems and was formally designed to "teach the staff of executives, managers and engineers principles for developing and installing efficient systems" for the following: quality control, production control, cost control, inventory control, plant layout, organizational control, methods engineering, pattern making control, and sample making control.

The users of the knowledge transmitted were primarily the chief executive officer and his production staff. However, the strong orientation of the volunteer toward organizational development effectively extended the user groups to supervisory staff and machine operators. The user group in fact became the entire plant. Some impact was felt in the sales office as well.

The situation triggering the change program was the need to rationalize the organization to improve quality and maintain markets. Improved internal organization was also considered necessary for securing any longer term technology support through licensing. The situation, then, was one of positive orientation to outside consultation, based on the past experience of the client, a critical problem of quality and loss of key staff, and the need to prepare for future relationships.

The change strategy of the volunteer called for the development of new systems by the management team working with the consultant. Emphasis was to be on understanding the "whys and wherefores needed to solve problems" and not just on how to resolve the existing complications. The strategy is easily identified as one of catalyzing capabilities within the firm, with prescription of techniques that accompany an overall organizational development activity. The tactics of the consultant's program rested on five basic devices:

1. Development of management principles with participation of the management group;
2. Comprehensive personnel training in every production area;
3. Weekly reports from the consultant to the management team, including recommendations and progress achieved in implementing production techniques;
4. Demonstration of new concepts and techniques on the plant floor by management team and consultant;
5. Preparation of a detailed report, including working guides and additional goals, for use by client after project termination.

While working in the plant, the volunteer encouraged management to approach their leadership roles in a new way. By arriving early at the plant and being available, the volunteer demonstrated by his own work in the plant management practices that the management was to follow to set an example for the other personnel. The volunteer worked closely with management and production personnel in devising solutions to the problems of the firm. He was not merely prescribing, but introducing new ways of identifying and attacking problems. When he introduced a new concept, he trained the plant supervisor and workers in its application.

Frank and explicit weekly reports from the volunteer were

distributed to the staff as a form of feedback on the application of new techniques. These weekly reports and the final report together represented a valuable history of the consultation and the ideas reviewed and implemented.

The combination of confrontation, catalyzation, and demonstration were reinforced by the continuous presence of the consultant in the plant. When some workers argued that they could not learn new ways of carrying out tasks, the consultant sat with them, forcing them to experiment with the new procedures. After repeated attempts, the production specialists agreed that their capacity for changing and modifying routines was greater than they had expected. This attention on the part of the consultant led to gradual change in the behavior of management as well. Several key staff doubted the usefulness of many of the volunteer's suggestions but eventually changed their opinion after relentless pressure by him to experiment with the new techniques. The cultural orientation to agree verbally with recommendations but to delay action because of lack of full commitment to change was overcome by the direct demonstration and insistent involvement of the consultant with the personnel concerned. These tactics, coupled with the continuous feedback of information from the consultant to management, reinforced the learning process.

The breadth of knowledge and experience of the consultant aided in the success of the project. Having written and taught in the fashion industry and managed similar activities for twenty-five years, the consultant was equipped not only to solve problems but also to teach the methods of problem solution and change.

The type of technology supplied can be primarily classified as controlling. Quality control, inventory control, production control, and production procedures were the primary thrust of the project. Yet the consultant also worked diligently at instilling a new orientation and effectiveness in the leadership. Emphasis was given to delegating authority and supporting staff decisions. Authority was formalized and responsibilities classified, in addition to providing new concepts of communication and motivation. Organizing technologies were present both in the development of the authority structure and in production flows. Planning was also affected since the consultant encouraged the integration of control information and company strategies into a formal plan to be reviewed periodically with key staff. The goals of the consultant covered all general areas of the organization except sales, which were moderately affected by the new organization and planning concepts.

Results of Project. The results of the consultation project were not immediately visible, except in the changes in the attitudes of management. Certain changes in control procedures and plant layout were instituted, but the more dramatic changes evolved at later dates. The diverse targets of the program almost overwhelmed the management, but the owner of the firm, after digesting the recommendations of the volunteer, became convinced of the long-term value of implementing more of the changes suggested by the consultant. Some six months after termination of the project, the owner asked that the same volunteer return to the company for several months. This was considered necessary to begin implementing the variety of changes that were needed in the firm; although the plans and goals were clearly identified, the presence of the consultant could help in effectively instituting these changes. In the second consultation, the original change program was put into action with greater commitment on the part of management. The senior management of Villa had not doubted the value of the changes, but they had underestimated the time, energy, and skill required to implement them in so many areas of the operations of the firm.

Eventually, the organization established new organization structures, expanded its planning, and instituted a number of control and production measures. Among these measures were:

- Quality control systems for all goods, from receipt of raw materials to shipment of finished garments;
- Production control system to keep minimum inventory in process and to track the location of all work bundles and their scheduled completion;
- Cost control to audit costs per lot produced;
- Inventory control to show the amounts, location, and storage time of all items of raw materials;
- New plant layout using minimum space and planning intratransport efficiency, taking into account seasonal changes due to style changes;
- Methods engineering to evaluate efficiency;
- Pattern and sample making controls for producing new samples for sales.

The confidence gained in the control area apparently helped to provide a more receptive atmosphere for changes in the general areas of organization and planning. Had these areas been addressed individually, the project might not have resulted in such widespread change.

Immediately after the second consultation project, the client entered into a licensing agreement with a U.S. manufacturer. The U.S. firm supplied specifications, production manuals, and training materials and scheduled one to two week visits to the plant every three months during which certain improvements in operations were suggested, problems diagnosed, and special training carried out. The production manual supplied by the U.S. firm is used as a basic guide by the organization. The licensing agreement also provided new product opportunities, including shoes, bags, accessories, and other product lines that could be introduced in the future. The opportunities offered by the licensing arrangement confirmed the belief of the client in the value of the organizational changes that were implemented with the insistence of the volunteer consultant.

The economic savings that resulted from the new control activities encouraged change in other areas. Methods engineering alone reduced the number of trucks used to store and transport goods, improved sewing operations, and accounted for an estimated annual savings of $8,000. Other savings came from reducing the use of waistband interlining and converting available machinery. Overall, the direct savings from control procedures were some $46,000. The inventory improvements, quality control, and other management efficiencies should reap much larger economic benefits for the firm.

Having seen the benefits of instituting new procedures, Villa management became more committed to securing technology on a continuous basis. The long-term licensing agreement was negotiated with greater understanding of the training and systems needed and with improved efficiency to sustain and benefit from a long-term relationship with a foreign firm.

In this project the internal diffusion of technology was substantial. The consultant involved almost the entire plant in activities of problem identification and solution. With his weekly feedback reports, he built up a dynamic change process, which was reinforced daily through demonstration. The owner of Villa supported the broad-based involvement, which insured wide diffusion. The traditional proclivity of competitors to withhold information and the lack of local diffusion networks inhibited any significant external diffusion of the technology. Since the changes resulted in no dramatic product or sales changes, external demonstration is limited. The management has encouraged its friends to investigate using foreign consultants. However, their lack of experience in using foreign technologies prevents them from readily identifying the benefits associated with such contact.

The forces working against change in the firm were numerous.

Though the owner had a positive orientation to new management concepts, he was unprepared for the scope and difficulty of change. The time between the two consultations allowed for the development of the commitment to change and an understanding of the requirements of its implementation.

The tendency to listen and agree, but not discuss, reservations about new systems would have hampered the effectiveness of the project had the consultant not pushed continuously for experimentation. Group participation in the consultation process, and a conscious effort to build up the confidence of an employee in his or her own abilities, were also critical factors in bringing about a real change in the behavior of the firm.

Finally, the local environment was so confusing and the atmosphere of distrust so prevalent among private sector managers that willingness to adapt and change was limited. In fact, some of the planning changes were not implemented because of the growing risk perception associated with the unpredictability of the general external environment.

In summary, widespread change took place because of the strategies and effectiveness of the consultant in addressing more than just organizing and controlling techniques. The emphasis on general organizational structure and personnel motivation patterns, in addition to system specific production processes, led to both a change in managerial practices and a solution of specific problems. The attitudes of the owner of Villa and his past experience with outside sources of technology gave a better understanding of the value of a consultant. The degree of change and the difficulty of change were not well understood by the client and he needed time to develop among his staff a stronger commitment to the implementation of new procedures and techniques. In this case, follow-up was important, perhaps essential, to effect real changes in the operations of the firm. As in many developing country firms, control changes had the most dramatic results in terms of immediate problem solution. However, a transfer of technology implies lasting change in leadership behavior. By concentrating first on controlling mechanisms, the consultant was able to break through to the more difficult area of leading, a strategy that resulted ultimately in greater personnel motivation and production efficiency.

PROCESS OF CONSULTATION

How consultants or other change agents implement new management systems has been the subject of much analysis and discussion.

Researchers have identified a number of major problems that may be encountered in a consultation process:[4]

- Mismatch between problem and solutions;
- Failure to deal adequately with power relations within a firm;
- (Perceived) threat due to the uncertainties of change;
- Unsatisfactory manager/consultant working relationship;
- Inadequate criteria for problem selection and solution evaluations;
- Environmental factors.

Diagnosis, or the matching of solutions to identified problems, has been seen as a critical factor in successful technology transfer. How the volunteer consultant deals with client attitudes, power relationships within the firm, and other factors is in part determined by his selection of strategy, his approach to the different stages of the consultation program, and the stage of development of the firm.

Technology or knowledge transfer involves several processes of information exchange, training, and learning. The way in which the consultant approaches the individuals in an organization to assist them in solving problems determines to a great extent the outcome of the project. Attitudes, motivations, capabilities, and strategies interact to make the process in which the client and consultant work together the most important overall determinant of the success or failure of technology transfer. Of the six factors determined in the study to be the most important in predicting success or failure, three relate to the process of consultation: the strategy for consultation; how the project is terminated; and the entry stage, which includes the consultant's diagnosis and formulation of work plan.

The analysis of a consulting project is based on the Kolb-Frohman model, which isolates five stages in a consultation:[5] (1) diagnosis (the explicit problems are identified for the consulting effort); (2) entry (the consultant establishes his working relationship with the client and may rediagnose or redefine the work plan); (3) planning-implementation (the consultant develops and begins the actual work program); (4) evaluation and termination (the client is left with the capability to use the new techniques introduced during the project and to seek assistance from outside sources in solving problems); and (5) maintenance (this is often possible only through ongoing contact with an outside consultant). The research of this study shows that those projects characterized by positive experiences in the entry and diagnosis stages tended to have a high ratio of successful technology transfers.

The volunteer's evaluations of the importance of twenty-seven components into which the five stages of a consulting project were divided indicate that the most positive factors of a consultation process are ease in determining how the work activity would be structured, maintenance of ongoing discussions during the project, and concentration on a combination of problems. The least positive factors were insufficient time spent by the client and consultant on isolating problem areas and the requirement for outside support for technology maintenance after project termination.

Table 3-15. Important Characteristics of Consultation Process (N = 472)

	Mean[a]	*Standard Deviation*[b]
	1.00 = Negative factor 3.00 = Neutral factor 5.00 = Positive factor	
Ongoing discussions were fruitful	4.2	0.9
No problem deciding leadership for work program	4.1	0.9
Consultant worked on several problems	4.0	1.0
Insufficient time spent diagnosing problems	2.6	1.1
Project requires outside support	2.7	1.2

[a]The mean is the overall average of the score on a scale from 1-5 with 1 being the least agreement, 5 being the highest degree of agreement, and 3 being neutral.

[b]The standard deviation is the degree to which answers cluster around the mean. The lower the standard deviation, the greater the clustering around the mean.

The examination of the process of consultation, or how the volunteer and client work together to solve a problem, supports the conclusions on the importance of diagnosis and general management development. When management is convinced of the value of change and technology improvement, there is a better chance of successful transfer. The volunteer's identification of the specific problems to be addressed in the consultation, in addition to the client's motivation and understanding of the value of the technology needed, are important elements in the early stages of a successful transfer of technology. The analysis of 472 volunteer questionnaires indicates that they believe that projects that focus on general management problems as well as specific production problems tend to be most successful.

What kinds of traits characterize the most effective consultant? International experience and previous consulting experience are apparently less important than general and specialized management know-how. The analysis of the 432 questionnaires shows that the consultants themselves consider their general skills to be as important to the consultation process as their system specific knowledge, a finding that reinforces the conviction that the most critical problem in developing country firms is the need to develop general management capabilities.

Table 3-16. Importance of Consultant Traits (N = 432)

	Mean	*Standard Deviation*
	3.0 = Neutral factor	
International experience	2.9	1.3
Prior consulting	3.5	1.2
Specialized skills	4.4	0.9
Generalized skills	4.4	0.8
Familiarity with client's country	2.6	1.2

The 432 evaluations by the volunteers of the relative importance of their management skills to the project coincides with the slightly higher emphasis they give to their role as problem identifiers as opposed to problem solvers.

Table 3-17. Role of Consultant (N = 432)

	Mean	*Standard Deviation*
	3 = Neutral factor	
Problem identifier	4.7	0.5
Problem solver	4.6	0.6
Planner	4.5	0.7
Teacher	4.4	0.8

The case of Hanil Yukan Hoesa in Seoul, Korea illustrates the concern of one volunteer consultant with identifying problems and teaching new skills instead of only with solving problems. A producer of carbon paper, the client firm recognized quality deficiencies in its products and sought outside assistance for correcting them. The client, which imports many of its raw materials, was interested primarily in a rapid solution to the quality problems. The volunteer consultant identified the source of the problem and worked with the firm's chief chemist in developing new formulae for paper production. A delay in imported raw materials precluded testing of the new formulae prior to the consultant's return to the United States. The client manager expressed great disappointment at the consultant's departure, feeling that the firm had not received the assistance it needed. When the raw materials did arrive, the firm tested the new formulae in production, only to find that they yielded no better results than had earlier formulae. However, because the volunteer had emphasized methods for identifying problems and generating new formulae in his work with the firm, the chief chemist was able to develop new formulae which, when tested, produced high quality carbon paper. Although this project did not result in a specific problem solution, it did equip the firm with skills needed for solving problems, and resulted in a transfer of technology. In our interviews with Hanil Yukan Hoesa, personnel expressed great satisfaction with the ultimate result of the project and gratitude to the volunteer for teaching staff chemists new techniques for solving problems.

The strategies and approaches used by the consultants evidence the concern with building internal management capability, in addition to providing solutions to specific problems. The volunteers find the prescriptive approach most useful in solving specific problems, but they also emphasize the importance of using more developmental approaches, such as theories and models, and attempts to catalyze staff into new activities in helping the firm to find new ways of dealing with problems. What they are least concerned about is helping the client merely to accept his situation.

Table 3-18. Consultant Strategies (N = 470)

	Mean	*Standard Deviation*
	3.0 = Neutral factor	
Helping accept situation	3.1	1.2
Catalyzing new efforts	4.1	0.7
Confronting problems	4.4	0.6
Prescribing solutions	4.6	0.6
Using theories and models	4.3	0.8

Catalyzing new efforts of client personnel was the strategy that had the highest ratio of successful technology transfer. This was followed closely by strategies that provide theories or models. The more passive strategies of confronting problems and helping the client to accept his situation had the lowest ratios of successful transfer.

Table 3-19. Consultant Strategies by Results of Technology Transfer (N = 67)

	Successful	*Neutral*	*Unsuccessful*	*Percent Success in Category*
Helping accept situation	7	3	5	47
Catalyzing new efforts	25	4	3	78
Confronting problems	16	7	1	67
Prescribing solutions	41	9	9	69
Prescribing theories or models	34	7	6	72

One volunteer consultant's strategy involved a preconsultation session with the client in the United States before traveling to Peru. In the meeting in the United States, the consultant and client discussed for several days the general principles and industry-specific techniques of the client's business, which is tourism and hotel construction, with little or no direct reference to the characteristics peculiar to the client's hotels. The two agreed on general principles of how a hotel should be designed and what services should be offered. The result of this preproject session was a frame of reference for the actual consultation. For example, disagreements during the project between the client and volunteers were resolved by referring to the general principles and outline of hotel design that had been articulated in their earlier meeting. These principles not only facilitated the consulting project and were the basis for solving problems in the client's hotel design, but also provided the client with a tool for designing hotels in the future and, as such, represented a successful transfer of technology.

The techniques used by the consultant reflect his commitment to a particular strategy of change. Though few of the consultants

Table 3-20. Consultant Techniques (N = 432)

	Mean	*Standard Deviation*
	3.0 = Neutral factor	
Human relations	4.7	0.7
Introducing new systems	4.5	0.9
Organizational behavior modifications	3.9	1.0
Changes in information use	3.9	0.9
Changes in tasks	3.8	1.1

interviewed in the study had sophisticated knowledge of organizational behavior mechanisms, they did believe strongly in the importance of human relations in working with clients.

Case Study: Sociedade Industrial de Equipamentos e Estruturas, S.A., Brasil (Rio de Janeiro, Brasil)

This case illustrates how environmental, organizational, and consultation factors can interact favorably to result in successful problem solution, internal technology transfer, and diffusion of managerial technology. The recipient organization, a family-controlled enterprise engaged in the manufacturing of a wide range of railroad equipment, agricultural machinery, and metallic structures, had been expanding and diversifying at an astonishingly rapid rate. Despite some significant environmental constraints, such as lack of qualified personnel and local technology, and fluctuating policies, Brazilian economic development and market environment were supportive of the growth of the firm. Although the company was market oriented and progressive, the rapid expansion of operations had put a severe strain on the relatively traditional organization structure and managerial procedures of the firm. Management was adaptable, resourceful, and familiar with searching, importing, absorbing, and adapting industry- and system-specific technology, and thus was receptive to the volunteer and willing to implement his recommendations for improvements in production organization and management.

SOCIN (Sociedade Industrial de Equipamentos e Estruturas, S.A.) is a family-controlled corporation engaged in the manufacturing of railroad equipment, metallic structures, and agricultural machinery. The company has two plants in the states of Minas Gerais and Rio de Janeiro, a central office in the city of Rio de Janeiro, and business offices in three other major Brazilian cities.

SOCIN was established in 1916 and initiated its activities with ore mining. Ten years later it had become a shop for repairing railway cars imported by the railroads, which at the time belonged to foreign companies. When the railroads were nationalized in 1946, SOCIN was technologically equipped to design and manufacture the first orders for passenger and freight cars. Because of its engineering and production know-how, the company has now gone beyond the domestic market and participated successfully in international tenders for railway stock, alone or through joint ventures. The growth of SOCIN in this line of products has been closely associated with the modernization of the government-owned railroads. More recently, within the framework of the governmental Railroad Development Plan for 1975-1979, SOCIN has been expanding its production of special freight cars for bulk transportation (including flat cars for containers and piggy-back trailers), containers, and passenger cars. In the meantime, it has also entered the market for electric unit-trains, self-propelled railroad cars, and subway trains.

The operations of SOCIN in the agricultural sector started in 1959 with the manufacture of implements for soil preparation and land clearing, such as disk-plowing harrows, land-clearing blades, tree-point hitches for big crawler tractors, and various tillage equipment. In 1968, it began producing harvesting combines. Today, SOCIN also manufactures forage harvesters, rice paddy combines, tractors and headers.

The development of hydroelectric plants in Brazil led SOCIN in 1960 to expand its activities in the lines of metallic structures. In 1967, it set up a modern unit for hot galvanization to produce relay towers for power lines, substations, and galvanized steel parts. In 1968, SOCIN added another field to its manufacturing operations: the production of steel pipes for basic sanitation works—water supply, sewage, and irrigation. In the broad category of metallic structures, and in addition to power transmission and to pipings, SOCIN also manufactures products such as dam gates, tanks, bridges, steel structures for hangers and industrial buildings, cranes, and ship deck machinery and equipment. The need to comply with technical standards and specifications for its own products led SOCIN to establish in 1973 a new line of items—cast parts in fiberglass reinforced plastic. In 1977 SOCIN began producing automobiles with fiberglass bodies developed by the firm and using some components (e.g., engines) supplied by a North American automobile company.

Railroad equipment (freight and passenger cars), steel pipes, and certain heavy steel structures and components are manufactured at the older plant in Minas Gerais. The assembly of electric trains, and the production of farm equipment, steel towers for high voltage transmission lines, naval equipment, molded parts in plastic reinforced with fiberglass and automobiles takes place at the newer and larger plant in the state of Rio de Janeiro.

The growth record of SOCIN is impressive. Before the 1974 consulting project, the company had more than doubled its sales revenues in equivalent current U.S. dollars, from roughly $13 million in 1971 to $30 million in 1973, representing over 50 percent compound annual growth. From 1974 to 1975 sales grew 82 percent, from $44 million to $80 million. This remarkable growth was achieved without a comparable increase in the work force, which went up only 6.2 percent from 1973 to 1974 (to 2,300 employees) and 13.5 percent from 1974 to 1976 (2,962 employees). Between 1975 and 1976 the number of employees remained practically unchanged.

Internal Firm Characteristics Affecting Technology Transfer. Task technology at SOCIN is both routine and nonroutine, depending on the product under consideration. Railroad equipment is usually produced to order on a routine or nonroutine basis and sold directly to the final customer. Most of the farm machinery, as well as automobiles, are produced to stock and sold to the consumer through a network of distributors. Metal structures, steel pipes, containers, and ship deck equipment are usually manufactured to order and supplied directly to the customer.

Except for some rearrangement in hierarchical levels, the basic organizational structure of SOCIN changed little from the time of the initial consulting project (1974) to the follow-up project (1977). Although the first consultation led to some improvements in the formal definition of responsibilities, the structure and job positions of the company had already been formalized in charts and descriptions. In 1977, authority was distributed through eight principal layers of managerial hierarchy versus seven in 1974. At the top, there was an executive board of directors composed of two directors—the president and the vice president—both of whom were members of the family that command a large majority holding in the firm. These directors were assisted in major policy formulation and decision making by a management committee—the so-called staff meeting—comprised of senior staff members and heads of divisions and departments. This committee met regularly once a month.

SOCIN was organized according to three broad functional categories. The marketing system had four divisions: (Railroad) Passenger Transportation Equipment; (Railroad) Freight Transportation Equipment; Agricultural Equipment, and Structures; Export Department. The production system encompassed two superintendencies, one for each of the two manufacturing plants (Minas Gerais and Rio de Janeiro). The administrative system included the departments of finance, controller, purchasing, legal, and administration. According to the general superintendent of one of the plants, there was inadequate coordination and communication among the various departments.

The process of personnel selection and admission varied with the position level. The decisions with respect to higher managerial positions usually fell to the executive board of directors, while the functional heads had jurisdiction over the selection of their respective personnel. In general, this process was coordinated and administered by the departments of industrial relations in each factory or by a personnel unit in the central office under the administration department. In filling positions, management tried to select people from within the company, but it also used employment firms for hiring specific types of employees, including engineers. Except for higher managerial levels, the decision to promote personnel within the firm was generally the responsibility of the functional heads, subject to the final approval of the corporate administration department, which reported directly to the executive board of directors. This process was informed by a formal personnel evaluation system.

In-company training was important to SOCIN, especially since its plants were located away from major urban centers where trained personnel or external training opportunities were readily available. Engineers were trained on the job and also sent to foreign manufacturers, such as SOCIN licensors, for study visits or traineeships. Plant workers were hired in the factory areas, generally without any previous formal training, and trained on the job. Those showing the greatest potential were chosen by the foremen to attend formal in-company training programs designed for lathe operators, machinists, and welders. These programs were not conducted on an ad hoc basis.

Although the strategic decision-making structure of SOCIN was dominated by the two directors, who belonged to the controlling family, it relied considerably on group participation. The senior management committee participated in major policy formulation and decision making. As a rule, the executive board of directors enlisted the cooperation of relevant division heads and staffs in the formulation of production and marketing policies. Since SOCIN was market oriented, production goals as well as certain production policies depended on marketing decisions. The commercial or marketing divisions proposed sales plans on the basis of market studies and sales forecasts.

While management in general had confidence in the staff and encouraged inputs from them, there was no formal mechanism to filter up the suggestions and concerns of employees. In principle, these were channeled through the various managerial levels. The primary employee motivating factors were monetary gain, job security, and organizational loyalty—all associated with working for a large and mature company—and acquisition of skills. According to one employee, a good number of the personnel had been with the company a long time. Salaries were competitive within the labor market, perhaps even somewhat above regional averages. Plant personnel at higher management levels were probably better off than they would be in big cities like Rio de Janeiro and Sao Paulo. Despite these positive factors, SOCIN has had difficulty in recruiting highly qualified personnel to work in its factories. Personnel are scarce, very much in demand, and often prefer to stay in major urban centers. SOCIN has been able, however, to retain trained personnel. This was illustrated by the relatively low turnover of technical and management personnel compared with the experience of other firms in Brazil. The scope of planning by SOCIN reflected the changeability in the external economic environment and in governmental development and credit policies. As one employee cited, "We are dealing with a very fluid situation. . . . We depend a great deal on government development policies and plans . . . and since government changes a lot, it is very difficult to plan." The high rate of inflation and cost of money also influenced SOCIN planning. Management tended to plan ahead for only one year, in part because the rapid expansion of the company made it difficult to do detailed planning for a longer period. A general budget or cash-flow plan was prepared, contemplating projections of sales revenues, current expenses, and capital investment outlays. While this budget was not very detailed, it was formally written and reviewed every month by the staff meeting or management committee on the basis of actual performance. The finance and controller's departments coordinated the program budget preparation and review processes, but the principal body for strategic planning and performance review was the management committee. The management policies and techniques of SOCIN were essentially the product of group thinking concerning the direction of the firm and routine accounting and cost data. Computerized data processing was only recently implemented to serve areas such as personnel administration and inventory management and control.

Production standards and procedures were set by top management and relevant staff, in particular from the plant engineering, production planning, and product engineering departments in each plant. A similar pattern prevailed with regard to office standards and procedures, though they were less formalized. Quality was monitored by the quality control department of each plant. At the time of the first consulting project in 1974, there was no standard cost system. The cost department analyzed cost variations of different production batches or orders by checking those costs against historical records. Standard costing was introduced recently by the newly created manufacturing engineering department. Budgetary controls were the primary responsibility of the controller's department, which as noted above, prepared monthly performance reports for analysis by the management committee.

With the exception of railroad car technology, much of the system specific technology of SOCIN has been imported, which is understandable given the level of Brazilian industrial development. SOCIN used Brazilian engineers to develop its own technical capabilities to manufacture railroad passenger cars and most types of freight cars, instead of resorting to license agreements with foreign firms. In the mid 1960s, it participated with a group of Brazilian manufacturers of railroad equipment in the development of rapid transit and commuter electric trains for suburban use. More recently, however, SOCIN concluded a license contract with a major West German concern to produce subway and presubway cars, streetcars and electric trains. Similarly, special freight wagons were made under a license agreement with a foreign manufacturer. A West German licensor was responsible for the production of containers as well. Naval equipment was produced under licenses from French companies. License contracts with two U.S. companies for agricultural equipment, such as the disc plowing harrows, other tillage equipment, and the self-propelled harvesting combine, have expired and SOCIN is now operating basically on its own in this area. Not only has the company absorbed the technology, but it is also introducing new technical developments in this product line. In automobile manufacturing, SOCIN uses some components from a U.S. company.

In addition to direct technical assistance from licensors and equipment suppliers, SOCIN engineers have visited these companies or done traineeships with them. Other than this, SOCIN personnel have not had any formal training abroad. Another source of technical and managerial know-how is represented by foreigners who have worked for SOCIN. The company employs a considerable number of people from Argentina, Italy, Spain, and Hungary. Two of them hold top managerial positions; the others perform managerial functions at the manufacturing plants. The principal reasons for recruiting foreign personnel are the rapid growth of the firm and the lack of locally available qualified personnel to meet the growth needs. Except for the two volunteer projects discussed here, SOCIN has not received know-how from foreign consultants.

Given its fast expansion, SOCIN has encountered some difficulties in finding, acquiring, and adapting technology, although in general it has been able to overcome them. These difficulties stem primarily from the following factors, most of which are related to the developing stage of the Brazilian economy:

- Generally insufficient level of Brazilian technological development to provide system specific technology or information. This situation has also been aggravated by the location of SOCIN plants outside of the major urban areas.
- Shortage of qualified personnel, particularly middle level technicians and middle management.
- Insufficient market size to justify research and development expenditures in new products. For this reason, SOCIN had to limit its research and development investments to adapting products to Brazilian conditions.
- Dependence of SOCIN operations on government development and technology import policies.
- High inflation rate, high cost of money, and relatively high short-term indebtedness.

External Environment Factors Affecting Technology Transfer. Although extremely favorable to the expansion of SOCIN, the Brazilian economic environment has been changeable due in part to the multiplicity of opportunities resulting from fast economic growth and structural shifts and to governmental development policies. In addition, SOCIN has faced shortage of some materials and delays from suppliers. SOCIN has been resourceful and responsive in adapting to this volatile situation by capitalizing on new opportunities and by expanding and diversifying its product lines. Despite stiff competition, particularly in the markets for farm equipment and metal structures, SOCIN has managed to grow at an extraordinary pace and to capture an increasingly larger share of the domestic market for its products. In the railroad equipment market, SOCIN holds 20 to 25 percent of the total production capacity of the Brazilian industry; in the agricultural equipment sector its share of the market is between 15 and 20 percent.

Government policies and regulations have a great impact on the operations of SOCIN. The firm depends heavily on government development policies and plans, especially with respect to railroad equipment and certain metallic structures for which the principal customers of SOCIN are government enterprises. These enterprises operate in the context of national development plans as well as sectoral plans such as the Railroad Development Plan, the National Sanitation Plan (PLANASA), and other plans dealing with electrification or electric energy distribution, hydroelectric power plants, and shipbuilding. Sales of farm equipment are closely related to government agricultural policies and incentives; in particular, they depend on the financing to farmers that is normally granted by government-owned development banks. In general, SOCIN activities are strongly conditioned by the government development financing policy with regard to capital goods. This credit policy is carried out by state and regional development banks, which are part of the BNDE System (National Economic Development Bank), in compliance with planning policies and directives from SEPLAN (Planning Secretariat of the Presidency of the Republic). In addition, SOCIN prices are controlled by a government board (Interministerial Price Council). With a few exceptions, manufacturers and traders wishing to increase their sales prices must seek prior authorization from IPC. In those cases where they are permitted to raise prices without prior authorization, the increases are subject to the filing of cost information justifying them.

Since SOCIN has relied extensively on foreign technological inputs, primarily in the form of license agreements, the company has fallen under the jurisdiction of government regulations pertaining to importation and transfer of royalties (in particular, the so-called *Ato Normativo No. 15* (Normative Act No. 15) of September 11, 1975, dealing specifically with transfer of technology). The bureaucratic processes of such governmental agencies as the INPI (Instituto Nacional de Propriedade Industrial or National Institute of Industrial Property)—an agency that is responsible for the registration of patents and trademarks that are owned by foreigners and for the approval of agreements that involve the utilization of foreign technical assistance by Brazilian entities—the Central Bank, and the Bank of Brazil are burdensome, often resulting in undue delays in project approval. However, according to one SOCIN employee, "in spite of the red tape, SOCIN has been able in general to get what it wanted."

SOCIN has, to some extent, benefited from governmental tax and credit incentives applicable to industrial development, exports and in-company training. The eligibility for and allocation of industrial development incentives are determined by the CDI (Council of Industrial Development). These incentives include the reduction or exemption of import duties and the federal excise tax on equipment, machinery, and components imported or manufactured in Brazil, eligibility for local development financing from official credit institutions, and accelerated depreciation of Brazilian-made fixed assets. SOCIN has been exempt from paying the INPI on the production of agricultural equipment. In addition, when the company set up its second plant in the state of Rio de Janeiro, the local city hall arranged land for SOCIN at cheaper prices and exempted it from all city taxes. Export incentives encompass a deduction from corporate taxable income equivalent to the proportion of gross sales to total exports, exemption from the federal excise tax, exemption from the state sales tax, credits on both the excise and sales tax equivalent to the excise tax that would have been due had the goods been sold on the domestic market, and eligibility for local short- and long-term financing of exports from government sources at favorable terms. Since SOCIN exports railroad and agricultural equipment, principally to countries of the Latin American Free Trade Association (LAFTA), it has been taking advantage of these benefits. SOCIN is also availing itself of the tax incentive for professional training of personnel, enacted at the end of 1976, through which it is entitled to write off from its taxable income training expenditures, provided the training programs have the prior approval of the Ministry of Labor.

With regard to the local scientific and technological infrastructure, the areas in which the two plants of SOCIN are located do not in general supply skilled labor, or offer university training. Universities located outside these areas in Brazil do provide reasonably adequate primary training for SOCIN needs in the basic field of mechanical engineering. It is impractical for SOCIN to release management personnel for extended periods of time to attend managerial technical tariff programs that are offered in major cities. "If one is operating in smaller cities, there is not much that can be done (regarding specialized training)," argued one employee. According to him, the major personnel problems in the Brazil of today is not so much the lack of university-graduated engineers as the lack of middle level managers and technicians. "There is a general shortage of vocational schools in the country. . . . In Brazil everybody wants to be a doctor. All the efforts (by preuniversity students) are spent in trying to pass the admission test to the university . . . and, when they finally graduate, they frequently come out without any real specialization."

Other local sources of managerial and technical know-how were generally considered inadequate. Contacts with industry and professional associations have not been significant. There has been some use of local consultants in the areas of administrative organization and personnel evaluation. Contacts with competitors have been restricted to the exchange of market information and no interchange of know-how of management techniques has taken place.

As mentioned previously, SOCIN has encountered difficulties in recruiting skilled employees, though not in retaining them. This, however, is due more to the general lack of qualified personnel in Brazil than to the specific location of SOCIN factories.

The local physical infrastructure has not posed any special problems for SOCIN since its plants are located near two important railroad junctions in the most industrialized area of Brazil and on a major highway connecting Rio de Janeiro and Belo Horizonte.

Process of Consultation. The first consulting project in 1974 was requested for the Minas Gerais plant and was designed to focus especially on its farm machinery products. According to the volunteer, SOCIN was expanding so rapidly that its organization could barely keep up with day-to-day problems, let alone plan for the future. Major decisions were based on enlightened guesses rather than on information developed from research. Management training and development were also badly needed. The combine harvester was the most sophisticated product manufactured, representing about 80 percent of plant production. Considerable increase in output was required in this product line. The fast growth of the company, coupled with the large increase in the parts manufactured by SOCIN and shortages of raw materials and delays from suppliers, put a strain on plant organization and uncovered weaknesses that, unless corrected, would seriously handicap company expansion plans. Recognizing these problems, SOCIN requested the assistance of a volunteer consultant who was familiar with the manufacturing of heavy farm machinery to:

- Review the entire production facility from the standpoint of organization, assembly line, production methods and controls, quality control and ways of coping with anticipated expansion;
- Make recommendations based on this review with respect to improved plant organization and layout, methods for determining machine loads or usage, improved production planning to insure timely acquisition of raw materials and parts, and personnel development and training programs.

The volunteer, who had spent ten years with one of the largest U.S. manufacturers of farm machinery and earth-moving equipment, stayed at SOCIN from July to October 1974. He worked closely with the plant general superintendent, with most of the plant department heads, and with several supervisors and foremen. The number of personnel that used the volunteer's know-how during and subsequent to the project was consequently large.

The volunteer's contributions covered a wide range of subjects. Specifically, the volunteer:

1. Assisted the plant's department managers in preparing new organization charts and job descriptions;
2. Recommended an organizational restructuring that reduced the number of departments reporting directly to the plant's general superintendent from seventeen to nine. The three major recommendations in the organizational area were to:

   - Form a manufacturing engineering department to handle all operation sheets, processes, methods, time studies, tooling, material handling problems, justification for new equipment and plant expansion. This department would enable company management to obtain essential information as to the feasibility and profitability of manufacturing proposed new products. Duties and functions were described, and operation and time study sheets were developed and implemented. These sheets were also to serve as the first step in establishing a standard hour incentive system.
   - Decentralize authority, in particular through creating the position of a production superintendent who would report to the general superintendent and coordinate departments of fabrication, assembly, welding and painting, and storage and material handling. This would reduce to nine the number of persons reporting directly to the general superintendent.
   - Centralize in one department the records on purchased parts and materials, which at the time were being handled by three separate departments.

3. Made a total of forty methods improvement suggestions, 80 percent of which were implemented during his stay. They included the proposal of a piece-of-work (standard hour) system which, if applied, would greatly reduce manpower on machines and assemblies, several methods for more efficient material handling, and the elimination of helpers from many jobs, thereby reducing costs.
4. Designed a new plant layout of the sheet metal press shop (in the combine fabricating department) providing for aisles so that material-handling containers could be moved freely on fork trucks instead of by overhead crane. Completed the layout for the building where harrows, terex truck bodies, and containers would be made.
5. Outlined an in-plant management or personnel development and training program, and furnished training materials.

In carrying out his assignment, the volunteer applied a mix of change strategies. Although he provided SOCIN management with certain prescriptions and principles and models contained in the interim and final reports, he also emphasized the catalyzing and confrontational approaches to induce problem-solving capabilities, facilitate problem solution, and generate alternatives. He resorted to a combination of four approaches to information delivery: measure and feedback (critical observation of various production operations); explicit demonstration (process or operation sheets, time studies and training); discussions with the general superintendent, plant managers and other plant personnel; and explicit plans and models (organization charts and job descriptions, plant layouts, welding wire feed charts, and training materials).

The project was concerned with improving methods of organization and rationalization of plant structure, production process, and production management. Thus, the type of management technology addressed by the project was a mix of organizing, controlling and leading.

Results of Project. The results and changes attributable to the initial project were remarkable. During the time between the 1974 project and the follow-up project by the same volunteer in 1977, SOCIN expanded its product lines extensively. At the time of the 1977 project the company was in the process of launching a number of additional products.

With regard to problem solution and internal transfer and diffusion of technology, the volunteer noted in the final report of the 1977 project that the majority of the proposals made by him in 1974 had been adopted in part, if not entirely. The volunteer expressed his satisfaction in the following way: "Such a large percentage of suggestions on the 1974 report has been adopted that I feel that the organization is a progressive one; that is, they are receptive to suggestions which they can put into effect."[6] In general, therefore, the project contributed to the strengthening of the organization along technical and supervisory lines, enabling it to meet expansion needs more effectively.

This favorable account was confirmed by SOCIN management. As a consequence of improved production procedures and increased efficiency, substantial man-hour and cost savings had been realized. The number of employees increased only moderately, while production and sales (particularly of harvesting combines) had gone up considerably. By 1976, costs had decreased by approximately 20 percent as a result of more efficient methods. The volunteer was viewed as having a positive impact on personnel attitudes, performance, ongoing problem-solving capabilities and technology search behavior. "There is a tendency for personnel to settle into a routine and not progress. When an outsider comes in who is competent and relates well, they try to perform differently and better and to demonstrate their capabilities. And the superiors who had already gotten accustomed to lower performances are now in position to require more."[7]

While there was no external diffusion of technology relating to SOCIN operations, there was spillover in terms of affecting the search behavior of other local firms. As one SOCIN employee explained:

> In a small, provincial and little-innovating community like ours, which usually does not receive foreigners, a (foreign) consulting project may always have propagation effects, provided that the consultant is well chosen. He can create a favorable image if he is competent and possesses the appropriate interpersonal traits. . . . The IESC volunteer's recommendations remained inside the company, but the project stimulated the curiosity of other firms. The community became aware of the existence of services such as the IESC. This is also a way of opening a closed and provincial community to technical contributions from the outside. . . . Several executives (from other firms) have asked me about IESC, whether it is worthwhile to use its services and if we were pleased with them.[8]

(As a result of these contacts, a local biscuit factory initiated an IESC project in 1977.)

The value of the contributions of the project to the productive sector and national development can only be stated in hypothetical terms. However, increased efficiency and production of SOCIN benefitted agricultural development.

Besides the external environmental obstacles to technology search and acquisition, the only force working against change was the rapid expansion that SOCIN was experiencing, which demanded a great deal of management attention. Apparently this was not a serious impediment to the project because the plant management team was extremely eager and receptive to the volunteer's work.

NOTES

1. Daniel M. Schydlowsky and Juan J. Wicht, "The Anatomy of an Economic Failure: Peru 1968-78," *The Peruvian Experiment Revisited*, ed. A. Lowenthal.
2. William H. Newman; Charles E. Summer; and E. Kirby Warren, *The Process of Management*, 4th ed. (Englewood Cliffs, N.J.: Prentice-Hall, Inc., 1977).

3. The term "values" refers to the attitudes and orientation of management and other personnel regarding, for example, work ethic, group loyalty, pride of workmanship.

4. Michael Jay Ginzberg, "A Process Approach to Management Science Implementation" (Doctoral dissertation, Massachusetts Institute of Technology, Cambridge, 1975).

5. David A. Kolb and Alan L. Frohman, "An Organization Development Approach to Consulting," *Sloan Management Review* 12, no. 1 (1970):51-65.

6. From final report of volunteer submitted per IESC procedure to IESC and to SOCIN management.

7. From interview with SOCIN management by Carlos A. Horta.

8. Ibid.

❋ *Chapter Four*

Policy Considerations

The political debate on transfer of technology has thus far failed to establish an agreed frame of reference in which the developed and developing nations can deal productively with the impediments to scientific and technological development. It is well established that it is in the interest of all parties concerned to help strengthen the capability of developing countries to receive and utilize technology. Yet there is considerable difference of opinion, and misunderstanding, concerning the most effective and appropriate mechanisms for accomplishing this. Many of the national and intergovernmental programs that attempt to deal with the issues of technology transfer ignore certain basic problems of the ultimate technology recipient. The findings of this study argue strongly for restructuring development assistance programs to recognize the user firm as a critical component of development. This restructuring must involve looking at the stage of development of user firms in order to assess realistically his capabilities for determining technology needs and for receiving technology.

This chapter first reviews the concerns of developing and developed countries regarding science and technology, as expressed in various policy initiatives and action programs. Development assistance programs have for the most part been aimed at developing science and technology infrastructures and improving the capabilities of developing countries to negotiate with the technology suppliers. It is the contention of this book that there must be more programs designed to build user firm capabilities to receive and utilize technology if these other programs are to be effective. Toward that end are presented several examples of existing successful programs directed at the user firm. The final part of this chapter describes characteristics that should be considered in developing new programs, and proposes the cooperative industries program as a possible initiative.

CURRENT PROGRAMS

Developing Country Initiatives

The objectives of developing countries that have been discussed in Chapter 2 are reflected in national and intergovernmental initiatives. Many developing country governments have established national offices to formulate science and technology policies and to control and direct technology importation and utilization. Generally, national policies have focused on six major initiatives:

1. Improving technical infrastructures in order to encourage more effective technology exploitation and development;
2. Creating specialized training centers to help build specialized manpower suitable for technological development;
3. Creating specialized information and technical assistance agencies at the private and public level to improve the ability of the local firm to select and bargain for technology;
4. Establishing mechanisms to encourage local research, development, and engineering activities (this includes the desire to have more product development carried at the local level);
5. Developing specialized programs to encourage skilled nationals to return to their home countries;
6. Establishing government policies to control and direct technology importation and utilization.

Examples of these initiatives are

- National policy centers, such as CONACYT (Consejo Nacional de Ciencia y Tecnologia) (Mexico);
- Technical training institutes, such as ITINTEC (Instituto de Investigacion Tecnologica Industrial y de Normas Tecnicas) (Peru);
- Business associations, such as Korea Traders Association;
- Business schools, such as ESAN (Escuela Superior de Administracion de Negocios) (Peru), Fundacion Getulio Vargas (Brazil), and University of Monterrey (Mexico);
- Managerial training programs, such as CEBRAE (Centro Brasileiro de Assistência Gerencial à Pequena e Média Empresa) (Brazil)

- Industrial research institutes, such as KIST (Korea Institute of Science and Technology) (Korea);
- National Technology transfer registries, such as SIEX (Superintendencia de Inversiones Extranjeras) (Venezuela);
- Centers to facilitate technology transfer, such as KORSTIC (Korea Science and Technology Information Center) (Korea).

Intergovernmental Initiatives

The developing countries, through a wide variety of intergovernmental initiatives, are also trying to establish a new technological order designed to increase their bargaining power for foreign technology while simultaneously building up their capability to generate and use technology. Through the United Nations Commission on Trade and Development, the United Nations Industrial Development Organization, and many other United Nations and regional institutions, a variety of initiatives are now being undertaken to:

1. Establish mechanisms for creating linkages between developing nations in order to develop more technical cooperation at the government level;
2. Encourage establishment of subregional organizations that will pool resources and information on technological assistance;
3. Develop intergovernmental information centers (such as the U.N. Centre of Transnational Corporations) to provide information to developing countries on selection and bargaining criteria for technology transfer;
4. Develop regional research and development centers;
5. Develop specialized assistance programs for newly independent nations;
6. Encourage developed nations to provide more untied development assistance for science and technology programs;
7. Establish international codes of conduct that will create more favorable conditions for transfer of technology to the developing countries and help protect the developing nations from potential abuses by technology suppliers.

Specific types of examples of these initiatives are

- International Bank for Reconstruction and Development (World Bank) programs;
- Inter-American Development Bank (IDB) programs;
- Organization of American States (OAS) Pilot Project proposal;

- United Nations Centre on Transnational Corporations;
- United Nations Development Programme;
- United Nations Commission on Trade and Development;
- Business school associations, such as Consejo Latino Americano de Escuelas de Administracion (CLADEA).

These intergovernmental initiatives, which were restated most recently at the United Nations Conference on Technical Cooperation among Developing Nations (Argentina, September 1978), reiterate the emphasis of the developing nations on the need for intergovernmental and national governmental offices to control and direct technology transfer. The initiatives also indicate the continuing belief in the developing countries that the major issues for technology transfer and development are (1) assisting firms to search for and acquire technology; (2) developing local research, development, and engineering capabilities; and (3) developing national and intergovernmental assistance and control organizations.

The following table illustrates that the developing countries are taking a top-down approach to the development of technology assistance by focussing on building intergovernmental and national institutions. They fail to attack the problem of the inability of the user community effectively to respond to the initiatives and assistance being proferred or suggested by public assistance programs.

Table 4-1. Developing Country Initiatives and Organizational Targets

	Target		
Initiative	*Intergovernmental Organization Development*	*National Organization Development*	*User Firm Development*
Regional R+D+E	●	●	
Local R+D+E Centers	●		
Information Centers	●	●	
Policy Development Centers	●		
Model Laws			
Engineering Science Training		●	
Local Engineering and Consulting Firms		●	●
National and IGO Technical Associations	●	●	
Bargaining Assistance	●	●	●
National Policy Development		●	
International Codes	●		
	7	7	2

Little mention has been made of the explicit managerial deficiency at the user firm level except in special programs of such countries as Brazil and Korea. As seen in the following table, few activities are addressed principally at helping develop the managerial capability of the user firms to organize resources and direct planning efficiently. The design of these programs seems to reflect the concern of the developing world over bargaining power. The table recasts the programs in terms of different stages of a particular firm's technology development. It may be noted that most of the programs, if they affect individual firms, are aimed at helping them search, bargain, or acquire specific technologies. There is a conspicuous lack of attention to the primary stage of building the organization (organization development). It is not surprising to note the lack of success in some programs like the OAS pilot project and the Peruvian ITINTEC, which assumed the presence of a user community already prepared to search for technology. The initiatives cited here also seem to overlook the need to build management and therefore initiate planning and search behavior, before proceeding with developing better technology search, bargaining, or acquiring skills.

Table 4-2. Developing Country Initiatives and User Firm Development States

	Firm Development Stage			
Initiative	*ORGANIZATION DEVELOPMENT (Managerial)*	*SEARCH, BARGAIN & ACQUIRE*	*MODIFY AND MAINTAIN*	*R+D+E*
Regional R+D+E			●	●
Local R+D+E Centers			●	●
Information Centers		●		
Policy Development Centers		●		
Model Laws		●		
Engineering Science Training			●	●
Local Engineering and Consulting Firms	●		●	●
National and IGO Technical Associations		●		
Bargaining Assistance		●		
National Policy Development		●		
International Codes		●		
	1	7	4	4

Developed Country and U.S. Government Initiatives

Developed country governments and private groups have responded to the concerns of developing countries with a variety of program options. Examples of these initiatives are

- Programs through American education associations, such as American Assembly of Collegiate Schools of Business and the Institute for International Education;
- Cooperative institutes, such as Latin American Agribusiness Development Corporation S.A. (LAAD);
- Profit-oriented training courses through American Management Association;
- Volunteer counsulting programs, such as IESC, Canadian Executive Service Overseas, British Executive Service Overseas, Japanese International Management Association, Volunteers in International Technical Assistance;
- Private sources, such as Control Data Corporation through its Techno-Tech/World-Tech program;
- U.S. Agency for International Development;
- U.S. National Bureau of Standards;
- Discussion in U.S. government on an International Industrialization Institute;
- Plans for creation of the Institute for Scientific and Technological Cooperation;
- Foreign missions.

A 1978 study by the National Research Council, prepared for the United States Department of State, outlines ten basic areas of assistance for consideration by the U.S. government in organizing development assistance activities (Table 4-3). These programs do not differ significantly from existing national and intergovernmental assistance programs in terms of the target at which they are directed. As can be noted in Table 4–4, the heavy emphasis is in the areas of national government organizational development, and on programs to assist the user firm in bargaining for technology, maintaining technology that has been transferred, and developing manpower. Again, little attention is given to developing the managerial capabilities needed for strengthening the organization's capacity for accurate planning and diagnosis of its technology needs.

The most recent initiative of the United States government proposes some programs specifically targeted at user firm development. The Institute for Scientific and Technological Cooperation (ISTC) has been initiated by President Carter, and its planning document calls for programs that address different stages of the development of

Table 4-3. U.S. National Research Council Initiatives and Target Organizations

	Target		
Initiative	*Intergovernmental Organization Development*	*National Organization Development*	*User Firm Development*
Development Assistance	●	●	
Education Programs (including management)		●	
Scientist and Engineering Exchanges		●	
Productivity Centers	●	●	●
Small Industry and Rural Development		●	●
Research and Development Centers		●	
Technology Corps		●	
Research in Industrialization		●	
Search and Selection			●
Negotiation Aid			●
	2	8	4

Table 4-4. U.S. National Research Council Initiatives and Firm Development Stage

	Firm Development Stage			
Initiative	*ORGANIZATION DEVELOPMENT (Managerial)*	*SEARCH, BARGAIN AND ACQUIRE*	*MODIFY AND MAINTAIN*	*R+D+E*
Development Assistance				
Education Programs (including management)				●
Scientist and Engineering Exchanges			●	●
Productivity Centers			●	
Small Industry and Rural Development			●	
Research and Development Centers				●
Technology Corps	●		●	
Research in Industrialization		●	●	
Search and Selection		●		
Negotiation Aid		●		
	1	3	5	3

the user firm. This document cites consulting programs to assist firms in diagnosing problems as a method of building basic planning capabilities in user firms. In addition, a cooperative industries program is described that entails loaning corporate executives to developing country institutions to aid them in planning, general problem solving, and technology capability analysis. Other activities such as on-site management training and product development centers are also presented as ways of addressing the fundamental organization development problem.

ISTC also calls for initiatives that help improve marketing and technology networks, thus demonstrating recognition of the need for cooperation between local government and productive sector organizations. An area that has received heavy emphasis is assistance to firms in modifying and maintaining technology already acquired. The aspect of ISTC that is most appealing is its plan for a variety of initiatives that deal with the different stages of the development of a firm.

Table 4-5. U.S. Institute for Scientific and Technological Cooperation and User Firm Development Stages

	Firm Development Stage			
Initiative	*ORGANIZATION DEVELOPMENT (Managerial)*	*SEARCH, BARGAIN, AND ACQUIRE*	*MODIFY AND MAINTAIN*	*R+D+E*
Diagnostic and other Consultation	●	●	●	
Technology Banks		●	●	●
Rural Village and Community Development				
Cooperative Industries Program	●		●	
Functional Mobile Training				
On-site Management	●		●	
Linking Arrangements		●		
Extension Networks and Institutions			●	
Product Development Centers	●		●	●
Common Marketing Centers		●		
	4	4	6	2

THE NEED TO REFOCUS PROGRAM INITIATIVES

It is clear from this discussion of existing and proposed programs to promote development that these initiatives for the most part neglect the managerial needs of the technology recipient. Of those programs that are designed to assist the individual firm, most do not focus on the organizational development of the firm, which, as identified in this study, is a fundamental step in building technological capability. Instead, the programs concentrate on research, development, and engineering, and on providing information intended for use by the firm in selecting among alternate sources of technology. All of these activities assume that the user firm is prepared to search for and effectively implement technology.

Several programs have been particularly successful in developing managerial capabilities of local enterprises for diagnosing technology needs and using foreign technology. In this section, three of these programs are reviewed: CEBRAE, KIST, and IESC. The strength of these programs lies primarily in their focus on the technology user and in their emphasis on the development of the managerial capabilities necessary for effective absorption and adaptation of technology. All of these programs are associated with public and private information and technical assistance institutions; however, this association derives from a recognition of the value of their services in complementing the development of managerial skills in user organizations. These programs are founded on the assumption that a critical issue for improving technology transfer is the development of a user community that actively seeks and applies new techniques. The development of this user community, however, is contingent upon strengthening the capabilities of local firms to manage technology—to diagnose needs, to devise solutions, and to seek assistance in implementing solutions.

KIST, CEBRAE, and IESC are examples of programs established by developing and developed country government and private groups. All three reflect a conviction that improved technology transfer is not the exclusive responsibility of public or private organizations, of suppliers or users, but instead can best be achieved through the combined efforts of all. All three provide assistance that has practical application in local firms. Because of this focus on the needs of the technology user, all three are sources of information on technology receipt and utilization that could be useful to policy makers in the formulation of technology transfer and development policies.

CEBRAE (Centro Brasileiro de Assistência Gerencial à Pequena e Média Empresa or Brazilian Center of Management Assistance to Small and Medium Firms) (Rio de Janeiro, Brazil)

CEBRAE is the managing nucleus of the Brazilian system for providing management assistance to small and medium-sized firms

(SMFs). Operating formally under the auspices of SEPLAN, the planning secretariat of the presidency of the republic, CEBRAE is a nonprofit organization with a private law status. Funds for CEBRAE programs come largely from governmental agencies (SEPLAN, the Presidential Planning Secretariat, and BNDE, the National Economic Development Bank). Other sponsors, which are critical in supporting CEBRAE programs, are universities and other educational and research institutions; employers' organizations; national, regional and state development banks, and planning agencies; and private consulting and research organizations. CEBRAE was established in 1972 under the sponsorship of the then Ministry of Planning and General Coordination (today's SEPLAN) by BNDE (National Economic Development Bank), FINEP (Agency for the Financing of Studies and Projects), and ABDE (Brazilian Association of Development Banks). The aim of CEBRAE is to contribute to the improvement and modernization of the SMFs in all business sectors with respect to technology organization and use, organizational design, and management techniques and procedures. The organizational structure of CEBRAE allows for focusing on small and medium-sized enterprises throughout Brazil and tailoring programs to their needs. CEBRAE works through agents called CEAGs (Centers for Managerial Assistance), which are active in twenty-two states. The CEAGs conduct sectoral and regional studies to identify user firm needs, provide specific technical and managerial assistance to SMFs, conduct training programs for top and middle management and for consultants, and support SMFs in preparing financing and loan applications to development banks. CEBRAE directly manages training and development programs for CEBRAE personnel; maintains a bank of case studies on management; conducts special projects for assistance to firms with no more than ten employees; manages assistance programs to small and medium-sized export firms; provides support for joint ventures by SMFs, such as common sales organizations and accounting services; administers the national productivity campaign launched by the Brazilian government; and sponsors management training programs for development banks.

CEBRAE uses financial resources from SEPLAN and BNDE to support the CEAGs programs, contributing 55 percent of each CEAG budget. The remaining 45 percent is covered by contributions from state development banks, other CEAG members, and firms that have been assisted by CEBRAE.

CEBRAE technical and managerial assistance to local firms has increased dramatically since its founding. In 1973, 1,077 firms received assistance from CEBRAE. By 1976, that number had grown to 5,795. The number of participants in the management training program of the system tripled from 12,535 in 1973 to 36,963 in

1976. CEBRAE received special recognition in 1975 in the Second National Development Plan (1975-1979), which identified the system as a fundamental instrument of national development policy and ascribed to CEBRAE specific responsibilities to discharge in the area of SMFs.

CEBRAE has been effective in transferring and diffusing managerial technology because of its focus on the technology user and on mechanisms for increasing the abilities of the local organizations to use new technologies. CEBRAE conducts research to develop information that can be applied practically by SMFs in their operations or by CEBRAE in developing new assistance programs. Much of this information results from CEBRAE contact with foreign sources of technology and from careful monitoring of programs in other countries designed to assist managers in small and medium-sized firms. Through contact with, for example, the U.S. Small Business Administration and similar organizations in Europe and Latin America, CEBRAE is able to keep abreast and adapt new programs to fit the needs of SMFs in Brazil.

Because of its knowledge of problems in SMFs and of programs that have been successful in other countries in improving local managerial capabilities to utilize foreign technology, CEBRAE is viewed by the Brazilian government as a valuable source of information on small enterprises. A report completed by SEPLAN in 1977 draws from the CEBRAE experience and outlines the Special Program to Assist the National Small and Medium Firms that has been approved by the president. At present, CEBRAE advises agencies of the federal government on policies addressing local business development and SMFs.

CEBRAE is a model for improving technology receipt and use by the Third World that may be applied in other countries. The focus of CEBRAE on assessing local firm managerial capabilities and deficiencies and on designing programs to develop those capabilities produces detailed information on the needs of the Brazilian business community. This information can be, and appears to be, delivered to policymakers and, in the case of Brazil, has contributed to policies on technology transfer that have practical applicability for local businesses and that address specific user needs.

KIST (Korea Institute of Science and Technology) and Other Korean Institutions (Seoul, Korea)

The plans for economic development prepared by the government of Korea in the early 1960s were dependent for their success not only on the availability of capital but also on acquisition and utilization of modern technology by Korean business. There was a shortage of highly skilled manpower and laboratories for research and development. In most cases, production volumes were insufficient to justify the large investment of funds required for original product development or even the perfecting of adaptive technologies. The government, in recognizing that science and technology are integral components of development processes, created during the first two five-year plans the following institutions keyed to developing Korea's technological infrastructure:

1962: KORSTIC (Korean Science and Technology Information Center);
1966: KIST (Korean Institute of Science and Technology);
1967: MOST (Ministry of Science and Technology);
1970: KAIS (Korean Advanced Institute of Science);
ADD (Agency for Defense Development);
1971: KDI (Korean Development Institute).

KORSTIC, KIST, KAIS, and KDI were established as nonprofit organizations to develop programs to facilitate technology receipt, utilization, and diffusion by Korean enterprises. The government role in conceiving and financing the institutes was critical. However, the government also recognized that the technological infrastructure could not be developed rapidly through government institutions that do not involve the local private sector. As a result, KIST, KAIS, KORSTIC, and KDI operate as autonomous institutions, whose independence derives from their self-perpetuating boards of directors, from special laws that allow outside rather than government audit, and from endowments. These institutions are thus characterized by private/public collaboration, by a focus on the local private sector as well as on public programs, and by programs aimed at providing the local private sector with information that has practical applicability to its operations. Established in 1966, KIST is one of the oldest of these institutions and has been particularly successful in linking its research activities to private firms.

The aim of KIST is to conduct

> comprehensive laboratory investigations for applied and development research directly linked to Korean industry, as well as objective-oriented basic research in the fields of sciences, engineering, and economics, along with technical services including economic and engineering feasibility studies, assistance in the importation and adaptation of technology, and provision of technical information.[2]

Growth in KIST operations is reflected in the growth in its annual operating budgets, which increased from $58,000 in 1967 to $6 million in 1973. Half of the KIST staff of 1,000 are professional scientists, many of whom were working in developed countries when KIST was started, but who returned to Korea to work at KIST.

Research in the forty-seven laboratories of KIST is organized in twelve research groups, which are polymer, chemical, environmental, electronics, and mechanical engineering; organic and applied chemistry; food and biotechnology, pilot plant, materials science, metallurgy and metallurgical engineering, and industrial economics. Its contract research for industry covered 1,600 projects by 1976, and accounted for $7.6 million or 60 percent of its income in 1977.

KIST also maintains a number of specialized centers for software development, foundry technology, precision machinery, electronic communication research, and technology transfer.

Through its technology transfer center, established in 1976, KIST provides consulting services to local industries in selecting and arranging technology transfer. The technology transfer center also analyzes, evaluates, accumulates, and disseminates information necessary for successful technology transfer.

Once technology has been transferred, the center assimilates it and adapts it for more effective use by local industries, and provides consulting services to firms that have imported technology and that need assistance in its adaptation. Through the promotion and assistance of the center in arranging for transferring technology to a group of firms, many local companies that would not under normal circumstances have access to the technology are able to acquire and use foreign techniques.

An important aspect of the technology transfer center is its role as an advisory body for the economic planning board of Korea. Recommendations are submitted by the center concerning board approval of proposed technology import and licensing agreements. Because KIST programs focus on the problems and needs of technology users, its recommendations to the government can assist in the development of policies that have practical benefits for local industries.

IESC—International Executive Service Corps (New York)

Founded in 1964, the International Executive Service Corps is a private volunteer organization that uses retired business executives as consultants to private firms and government agencies in the developing world. IESC is funded by U.S. AID, U.S. corporations, private foundations, and, in recent years, by foreign clients, whose contributions now comprise 40 percent of its budget.

The aim of IESC is to provide technical and managerial assistance to client firms in the developing world. A typical IESC consulting project results from a client request for assistance in solving a specific problem relating to production, general management, finance, marketing, or personnel. A volunteer consultant is located from among the 10,000 retired businessmen on the IESC register and is sent on an assignment lasting approximately three months. Through working with an individual assigned as his counterpart or with a small group in the client firm, the volunteer diagnoses the problem of the client, recommends and may assist in implementing specific steps for solving the problem, and attempts to develop new skills in the client firm for solving its problems in the future. In many cases, the volunteer returns for a follow-up project to assist the client in adjusting to new procedures or to focus on a new problem.

By providing assistance at a cost far below that charged by a traditional consulting firm, IESC makes available to clients expertise to which they would otherwise not have access. Volunteers can provide assistance in a variety of industries, ranging from craft industries such as silk dyeing to computer adaptations to planning petrochemical projects. IESC volunteers are generally respected by clients as businessmen whose experience spans several decades. The short-term nature of the projects militates against feelings of dependence on the volunteer or on sources of assistance that are not available locally.

IESC clients include local private firms, industry associations representing a number of firms, government enterprises and agencies, and government training institutes. The impact of IESC consultations on individual firms, particularly private firms, is often quite significant, resulting in increased profitability, critical training of management or a change in managerial practices. Projects with industry associations may not be designed to alter the general managerial systems of individual firms, but may introduce new techniques or stimulate changes in the industry that have a significant multiplier effect in individual firms or in other industries. Similarly, projects focusing on improving training programs of government or other public institutions can affect indirectly the management capabilities of individual user firms.

IESC has contributed significantly to the development of managerial capabilities in the developing world, and is one of the most valuable in the developed country assistance programs. The contribution of IESC to improved technology receipt and adaptation by developing country firms derives in part from the 10,000 retired businessmen on its roster, who represent a vast pool of general business skills, and of industry-, system-, and firm-specific knowledge. The early focus of IESC on the private sector in the developing world gives it long experience in diagnosing problems of developing country firms, and assessing needs for new technologies. Much IESC work in the public sector is with government programs designed to stimulate local production and to develop local private sector management capabilities. For example, IESC volunteers have worked with both KIST and CEBRAE.

Continued access to IESC through follow-up visits by volunteers, or, in many cases, through direct correspondence between the volunteer and client allows developing country managers to reinforce new managerial techniques and develop skills that were introduced by the volunteer. The value of the IESC program is the ability to target specific problems in firms in virtually any industry, to locate a consultant with the appropriate skills, and to structure a consulting visit through which the client is provided the assistance he needs at a price he can afford. In many cases, the IESC project opened the client to new management systems and triggered a change in attitude among client personnel that has led to more effective management through better internal problem diagnosis and design of solutions. The strengthening of local skills for diagnosing and solving problems reduces dependence on outside assistance, increases the ability of local firms to receive and utilize technology, and thereby facilitates transfer of technology.

COOPERATIVE INDUSTRIES PROGRAM—POSSIBLE INITIATIVE FOR MANAGEMENT DEVELOPMENT

In Chapter 3 factors that affect the transfer of technology to the developing world were examined and constraints on the ability of the user firm to seek, receive, and use technology identified. In this chapter, it has been argued that the majority of current assistance programs designed to facilitate technology transfer have failed to consider adequately these constraints. Consequently, these programs have proved less effective than would be possible had they been better tailored to needs of user firms and particularly to the constraints that impede their initial search for technology. These constraints relate specifically to

Internal Firm Characteristics
- weak capabilities to diagnose problems;
- little understanding of the value of technology for firm operations;
- limited financial and managerial resources for acquiring and using new technologies.

External Environment Constraints
- environmental instability;
- lack of reliable information.

Process of Consultation Constraints
- high cost of consultants;
- lack of readily accessible consulting services;
- need for consulting services for diagnosing problems and implementing solutions; that is, general managerial and technical assistance.

It is in the interest of all actors in the technology transfer process—technology suppliers, their national governments, technology users, their national governments—to keep these constraints in mind and refer to them in evaluating existing programs or designing new initiatives.

Knowing these constraints, it is possible to design assistance programs with characteristics that address these constraints on user firms. Such a program would focus on the internal firm environment in order to provide directly (1) the diagnostic services and (2) the assistance in implementing solutions to problems that the user needs. It is often difficult for managers in the developing countries to see the relevance of formal management training to their operations, and training institutes in local environments are consequently not exploited fully by user firm personnel. By focusing first on the user firm, such a program would assist managers in diagnosing problems and understanding the pertinence of technical or managerial training for those personnel who will be responsible for implementing solutions, and would induce them to use these training facilities more judiciously. Other characteristics of a program that would address user constraints include drawing on existing sources of information and technology in the local environment, such as multinational corporation subsidiary personnel, private volunteer organizations, and whatever local consulting services are available. By drawing on these sources, instead of importing consulting services from abroad, the cost of the program would be reduced, and the continuing and ready access to consulting assistance that the user firm urgently needs would be more easily guaranteed. The following describes the Cooperative Industries Program that was designed with these characteristics in order to address the major constraints on user firms in seeking and acquiring technology.

The Cooperative Industries Program would be sponsored by the U.S. government, in cooperation with technical institutes in each participating developing country. The program would concentrate on consultations and workshops to encourage the participation of corporate technicians from U.S. international companies who are already working in the participating developing country. These technicians would bring diverse background and skills capabilities to help local firms (in industries other than that of the subsidiary) to diagnose managerial or technology-related problems and then to assist them in beginning technology planning activities. The same corporate individuals would also participate in continuing training programs at the local technical institution. Through these training programs and consultations, materials would be developed to establish a training and consultative capacity within the developing country institution.

After initial diagnosis, planning, and training activities have been completed, the local firm would be put in contact with other firms in the United States that would volunteer to host training visits, exchange information, and otherwise assist the local firm. Developing country firms would thereby receive valuable assistance from U.S. subsidiaries operating in the country and be linked to other smaller firms in the United States for additional and product-specific support. The support of the local subsidiary might be extended through the participation of IESC or other private voluntary organizations.

The establishment of the Cooperative Industries Program would involve several basic steps:

1. Identify a coordinating organization that can recruit and develop support of U.S. multinational corporations. This organization would ensure that U.S. corporate manpower would be made available to the program in the developing country. The coordinating organization would also develop a network of small and medium-sized firms in the U.S. who would be willing to communicate with developing firms on a regular basis concerning organization programs, sources of equipment, and possible training exchanges.
2. Identify local technical institutions in a variety of countries to host the diagnostic and training activities. These institutes would also be responsible for developing new training materials and manpower development programs based on the experiences and teaching aids created in parallel with the consultation program.
3. Create a specialized training and research group that would aid the local institutions in analyzing and studying the diagnostic and planning capabilities of the volunteers. This group would also be

responsible for helping to design training programs that could further use the available volunteer in a systematic and efficient way within the local environment.

The Cooperative Industries Program is conceived as a vehicle for aiding developing company firms in their organizational development. It also provides a means by which U.S. companies can participate without committing extraordinary manpower and finances. The steps of a typical project will demonstrate how they participate:

1. An MNC technician from a U.S. foreign subsidiary would be made available for four to six days to work with the local firm in an industry other than that of the subsidiary. During this time he would do a general diagnosis of the operation and suggest alternative changes and improvements. He would also consult in the establishment of a plan for improving technology utilization in the firm.
2. The research group, working with the local technical institution, would analyze and prepare a written report on the process of diagnosis as part of its development of training material. The diagnostic activity would thus have been recorded and put into an ever-enlarging package of cases to aid in training in many different countries.
3. Once the local firm has completed its initial diagnosis and plan, it would then go to the training institution and the coordinating group for further assistance. For example, if it needs information about certain kinds of equipment or operating procedures, it would be put in contact with the U.S. domestic firm. This firm would provide assistance and suggestions to the firm on a continuing basis. This would allow the developing country firm to maintain the momentum of its initial technology diagnosis and plan. It might even send individuals to the U.S. firm for training programs if the overseas U.S. activity is not adequate for the particular need. This U.S. firm would operate in the same industry as the developing country firm.
4. The local training institution, after having cooperated in a number of diagnostic consultations, would assemble specialized training programs on common areas of technology problem-solving and planning. This program would be the basis for more specific training programs on topics such as product development or quality control. The local U.S. MNC volunteers would participate as guest lecturers in this program. Teaching aids and materials

from the training program and analyses of the diagnostic consultation would then be exchanged with other institutions in the developing world so as to multiply the impact of the individual diagnostic and training activities.

Motivation for Participation of U.S. Private Sector Organizations

U.S. corporate support arises from both practical business interests and the volunteer ethic. If the program is comprehensive enough and coordinated by the U.S. government, large international firms will contribute manpower and training materials as part of their public affairs activities. In most countries, these firms are already providing volunteers for training and ad hoc technical assistance activities. If the people are only involved for ten to fifteen working days per year, there is not a serious impact on the subsidiary. For the smaller firm in the United States, the project allows for participation in overseas activities without having to make serious travel, financial, managerial, or other resource commitments. If it is to host a small training visit, remuneration by either the coordinating office or the interested foreign user firm would cover the cost.

Finally, this Cooperative Industries Program, because of its ability to stimulate a technology search and utilization activity at the firm level, provides an important link to other development programs. For example, continuing support for the local firm might follow from other appropriate organizations such as IESC, Pan American Development Foundation, Technoserve, Inc., or Appropriate Technology International. The essential element of the program is that available resources can be directed at the first-stage barrier associated with triggering practical technology activities of individual organizations in the developing and underdeveloped countries.

Summary

A critical issue for improved technology transfer is the development of a user community that actively seeks and applies new techniques. The development of this user community, however, is possible only through a strengthening of the capabilities of local firms to manage technology—to diagnose needs, to devise solutions, and, if necessary, to seek assistance in implementing solutions.

The major finding of this study—that user firms in the developing world generally lack the diagnostic capabilities for assessing their needs for new technologies—has important policy implications that argue for a new emphasis in assistance programs. If a firm has not yet developed internal problem-solving and diagnostic capabilities and cannot plan, it will not be able fully to exploit the opportunities offered by many existing assistance programs. What is needed is a reshaping of the existing policies that assume that user firms are in a position to exploit, modify, and develop technologies to reflect the actual stage of user firm development and to propose programs that build the organizational capabilities to diagnose and solve problems. Only then will existing programs aimed at assisting firms to acquire, modify, and develop technology be effective.

Because of greater understanding of one of the fundamental obstacles to technology transfer, that is, the limited managerial capabilities of user firms to receive and use technology, there is at present a great opportunity for the U.S. government and developing country governments to shape assistance programs that can improve technology transfer. In designing new programs, however, policy makers should look to those existing programs that have focused on the user firm and that have strengthened user firm managerial capabilities. Expanding programs that have proved effective in equipping local firms to receive technology, or that draw on existing sources of assistance in developing country environments may be the method for using limited resources most efficiently to further facilitate technology transfer.

NOTES

1. *U.S. Science and Technology for Development: A Contribution to the U.N. Conference.* Prepared by National Research Council, National Academy of Science, National Academy of Engineering, and Institute of Medicine. Washington, D.C.: Department of State, 1978.
2. Korea Institute of Science and Technology, *Descriptive Brochure.* (Seoul: KIST) no date.

Appendix I: Environment and Technological Infrastructure Analyses

INTRODUCTION

Chapter 2 showed that contemporary management theories argue for viewing an individual firm as an organization whose operations are conditioned not only by internal structural characteristics but also by factors in the external environment. The delivery of goods or services by a firm may be affected as much by its interaction with the external environment as by its internal management capabilities. This open systems view of the firm implies that environmental factors must be examined in order to fully understand the behavior of the firm. The market and government incentives and restrictions emerged in Chapter 3 as important conditioners of the receipt and use of technology by a firm. These are, however, but two elements of the environment and technological infrastructure of a country that may affect the management processes and technology acquisition of a firm.

Through the field research and interviews with developing country managers, two major types of external environmental factors have emerged as conditioners of the behavior of a firm. One type refers to general economic and political factors, and the other relates specifically to the technological infrastructure and the availability of resources that enable a firm to exploit new technologies. These types are displayed on the following page.

General Economic and Political Factors

Incentives and control systems

- government regulations;
- market characteristics and competition;
- inflation patterns;
- balance of payments.

Political stability

- turbulence;
- predictability of change in environment.

Technical Infrastructure Factors

Direct suppliers of knowledge and manpower

- universities;
- other industries;
- government programs;
- foreign firms;
- immigration;
- specialized training institutions;
- equipment suppliers;
- licensors;
- private volunteer organizations.

Local support systems

- financial institutions;
- local consultants;
- special government assistance agencies;
- service companies such as labs, tool, and dye shops;
- physical infrastructure including roads, water power, communications systems.

Diffusion systems

- standards offices;
- patent offices;
- business and professional associations;
- universities and training organizations;
- government information offices;
- general publications;
- free press.

Economic and Political Factors

Incentives and Control Systems. Government incentive systems like those in Brazil and Peru encourage training of personnel both through use of government institutions and inhouse training programs. Other regulations may make it difficult or prohibitively expensive, as has occurred in Peru, to import technology. Governments may proscribe, as has the Brazilian government, deducting for tax purposes fees paid for joining foreign business associations, and thereby thwart attempts by local firms to gain information on foreign technology. Quotas on foreign technicians or controls over currency flows directly affect the ability of an individual firm to acquire technology from a foreign source.

The market is the key to inducing or inhibiting the firm in taking risks necessary for technological improvement. Research indicates that many firms see little reason to alter existing management systems or change technology use behavior unless those changes will improve their market position. Local competition may also affect technology transfer. In certain industries, competitors are exposed to technological innovations and developments through each other and are forced to take risks in order to keep up with the changing market patterns. Firms enjoying a secure market position may have no real incentives for investing in technological change and improvement.

The general economic environment also may encourage or discourage risking limited resources in new technology. For example, the rapid deterioration in the Peruvian balance of payments during the research discouraged many firms from committing to new programs involving long-term technological change. As foreign reserves continued to dwindle in Peru, the government stepped in to control all international travel, purchase of foreign equipment, and other business activities that might negatively affect the balance of payments. As a result, many of the firms that had begun technological programs tied to foreign assistance were negatively affected and long-term development of their programs was thwarted.

Political Stability. Political stability has a great impact on planning and risk evaluation of the individual firms. Political stability seems to relate less to the ideological position of the government of a country than to the predictability of change in the environment. In a country such as Tanzania, although technology is not readily available locally, firms are willing to take risks and invest in foreign technologies because they are confident that governmental policies will not change radically or rapidly and that they will be able to continue using foreign technologies. In countries such as Peru that evidence great political instability and turbulence, firms lack confidence in their ability to predict how the environment will change and are unable to evaluate options for improved performance and growth. Political stability and predictability of change in the environment are factors often as pervasive as those of competitive or market forces and are exceptionally important in conditioning the firm and its ability to adapt and plan future growth.

Technological Infrastructure Factors

Direct Suppliers. Having access to manpower and know-how supplies is a principal concern of any company. As a rule, firms look to local universities and training offered by public or private institution programs as a prime source of their personnel and management resources. Another source of manpower is the general business community—many firms hire critical personnel away from other firms when they are planning to improve technological capabilities. As a result, other industries and competitors become an important direct supplier of manpower and specific technological systems. In some cases, certain government programs provide the type of experience that will be useful to independent firms and they may also be a source of manpower. Firms often have access to sources of technology, both hardware and software, through the purchase of equipment and supplies, through licensing, or through technical assistance agreements.

The firm may look beyond its own environment to foreign sources of information and/or manpower. In some cases, favorable immigration laws facilitate hiring of foreign key manpower for specialized skills. Import regulations may also encourage firms to acquire equipment, licenses, or other types of technology from foreign sources. In some cases—Tanzania and Kenya are examples—private volunteer organizations represent a valuable source of foreign technology.

Support Systems. Support systems are often essential to the firm in acquiring and using technology. Banks, government offices, or other general business institutions are in many cases important sources for financing, counsel, and general information for aiding the firm in its technological development. Local consultants provide specialized manpower and system-specific know-how only periodically, primarily because of their limited expertise. Managers in all five countries examined in this study indicated that stronger local consulting systems would contribute significantly to the efficiency of their operations. Many of the clients interviewed draw on whatever local consultants there are, but also rely heavily on foreign consultants. Foreign consultants are prohibitively expensive in some cases, however.

Government commitment to developing specialized assistance organizations may be a significant aid in improving technology transfer. These organizations provide management consultants and specialized technicians for local firms, publish information on technology sources, and offer special seminars and training programs in general and specific management techniques. A notable example of this type of organization is CEBRAE in Brazil. Specialized service facilities such as laboratories and tool and dye shops are important for developing or reinforcing the capability of a firm to adapt to changing patterns in the environment.

The physical infrastructure is an element in the local support system that may have a great impact on the ability of a firm to grow. Without sufficient power or water supplies, communication systems, or transportation systems, a firm may not be able to implement plans for developing and adapting new technologies.

Diffusion Systems. Diffusion systems are necessary for linking different parts of the technological infrastructure. An active patent office serves indirectly to make more information available to inventors, businessmen, and government technicians. Standards bureaus and standards organizations are vital as guides to firms in developing new products or improving existing products and procedures.

Business and professional associations may encourage exchange of ideas and discussion of new developments, and in some instances, have pooled resources for financing of specialized studies or information gathering systems. The effectiveness of these associations as information sources varies greatly among developing countries and industries, however. In some instances, particularly in Kenya and Peru, these associations are used more as lobbying groups for dealing with the government. In business associations in Korea and Brazil, there seems to be greater exchange of industry- and system-specific information as well as discussion of political and general environmental trends.

Universities and training centers can stimulate diffusion of information and exchange of ideas. For example, the Korean Institute of Science and Technology encourages companies to develop new products and research programs. KIST works closely with universities and advanced schools of engineering and scientific training in Korea, such as the Korea Advanced Institute of Science, and therefore is an important link between the public and private sectors.

Local and foreign journals and general publications were cited repeatedly by local managers as sources of ideas and information on recent industry developments. A free press also seems to be an important aspect of the technological infrastructure. A number of managers indicated they rely heavily on the local press for information on political, economic, and social developments. Confidence in the local press' accuracy in reporting news will encourage firms to plan growth based on readily available information presented in newspapers, journals, and through radio and television.

Analysis of Five Environments

An analysis of the five countries examined in this study indicates that Brazil and Korea have rather extensive direct supplier and adequate support systems. Emphasis has been placed in both Brazil and Korea on developing new kinds of diffusion systems and the governments encourage standards development and exchange of information through professional and business associations. In Peru, Kenya, and Tanzania, there is less of a commitment to these kinds of support systems, a fact that may have to do with the size of the gross national product and the development stage of the country.

Much has been done in Korea to develop very specialized incentive systems to develop technological improvements in local firms. The incentive system in Brazil is also well developed, although recent growing concern with balance of payments issues has resulted in governmental actions to reduce loss of foreign currencies. This has in many cases thwarted technological developments by local firms. In contrast, neither Kenya nor Tanzania has formulated extensive programs to develop local firms and both have focused more, particularly through parastatal enterprises, on large government-sponsored programs for industrial and technological development.

In the case of Kenya, however, governmental development projects reflect more the lack of local private sector capability than an ideological commitment to government ownership, and programs to develop local firms are, as industrialization in Kenya progresses, now cited as priority areas of the government. The Peruvian environment, unlike either of these two groups, has since the mid 1970s been characterized by economic and social turbulence as a result of the political experiment launched in 1968 with the military takeover of government.

Peru represents the most turbulent environment of the five. Brazil shows some degree of turbulence because of its recent changes to mitigate balance of payment problems and reorientations toward local technological developments. Although Tanzania and Kenya have selected differing development models (the former favors state control of production and the latter favors combining private and public sector resources for production), government policies in both have remained fairly constant and have contributed to generally predictable and stable environments. Korea is a country where local business capabilities are sufficiently strong and government programs encouraging public and private sector cooperation highly developed, the result of which has been a predictable environment facilitating rapid growth in the local business community.

Table I-1. General Evaluation of Local Environments

	Brazil	*Korea*	*Peru*	*Kenya*	*Tanzania*
Incentives and control systems (strong/neutral/weak)	strong	strong	neutral	neutral	weak
Political instability (stable/moderately turbulent/turbulent)	moderately turbulent	stable	turbulent	stable	stable
Direct suppliers (strong/neutral/weak)	strong	strong	neutral	neutral	weak
Support systems (strong/neutral/weak)	strong	strong	neutral	weak	weak
Physical infrastructure (strong/neutral/weak)	strong	strong	strong	neutral	weak
Diffusion systems (strong/neutral/weak)	strong	strong	weak	weak	weak

The following are detailed descriptions of the environments and technological infrastructures of the five countries examined in this study. These analyses focus on the general economic and political situation of each country, particularly as they affect the productive sector and on the local technological infrastructure and sources of technology assistance for local firms. Other issues that are addressed are sources of foreign technology, cultural characteristics that have a large impact on local production, the physical infrastructure and the effect on business, and specific incentives and disincentives for development of local productive sectors.

KOREA

General Policies of Government

Geoeconomic Base. The government of the Republic of Korea administers about one half of the Korean peninsula—38,600 square miles, comparable with the area of Virginia or Portugal. Of this land, 23 percent is farmed and 67 percent is natural forest, leaving 10 percent for all other uses. Of the 1977 population of nearly thirty-six million, one half lived in urban areas (after the influx from rural areas in the 1960s) of whom seven and one-half million lived in the capital city of Seoul and two and one-half million in the main port of

Busan. Population growth has slowed from a high of 2.9 percent registered during the post-Korean War baby boom to 1.6 percent per annum. The population density of Korea is among the highest in the world.

Economically active population (64 percent of those over fourteen years of age) totaled over 14.8 million in mid 1977, of whom 3 percent were unemployed. Of the economically active, 6 to 8 million lived on farms and 8 million in nonfarm households. Forty-eight percent of females were economically active. During the ten years ending with 1975, the percentage of those employed in manufacturing and mining had gone up from 11 to 19 percent and in services from 31 to 35 percent, while the number of persons employed in agriculture, forestry and fishing declined from 58 to 46 percent of the total.

At the end of 1975, 2.2 million persons were employed in manufacturing. Of these, 680,000 worked in 23,304 registered small industries (defined as those having assets of less than $1 million and fewer than 300 employees) and 823,000 in 925 large companies. The annual wage was $807 per annum in the small industries and $1,089 in the large ones. By mid 1977, average employee earnings in manufacturing reached $1,732 per annum. Except for tungsten, limestone, and ocean fishing, the Republic of Korea is not well endowed with natural resources. It functions essentially as a processing nation importing most raw materials, relying on its hard-working work force to industrialize them, and exporting a major part of the production (more than one third of GNP in 1977).

History. The Korean peninsula was settled some 4,000 years ago by tribes originating in Central Asia. Later, Chinese colonies were established in the north. When recorded history began in 57 B.C., the northern half, together with Manchuria, formed part of the Koguro Kingdom while the southern half was divided between the Kingdoms of Silla and Paekche. In 668 A.D. the Silla Kingdom prevailed over the other two, unified the peninsula, and ushered in a Golden Age of cultural life inspired by Buddhism, which was carried from Korea on to Japan. The subsequent Koryo dynasty (918-1392) introduced the first civil service examination system, schools, tax laws, bronze coins, printing, and fine porcelain. The first seventy years of the last dynasty, Yi (1392-1910), brought along with Confucianism many scientific and technological advances. Among them were the twenty-eight (now twenty-four) letter phonetic Hangeul alphabet and movable metal printing type, which appeared at least fifty years before Gutenberg.

The simple (in comparison with previously used Chinese ideograms) phonetic alphabet and the early introduction of the printed word put culture and education within the reach of everyone. They are the basis of the very high literacy of Koreans and their high participation in education beyond grade school today. The Confucian influence introduced during the Yi dynasty is still felt in the Korean code of conduct, which is essentially formal and oriented toward family security. Japanese invasion began in the sixteenth century, displacing first the Chinese and then the Russian influence, and culminated in the 1910 treaty of annexation. The brutal colonial subjugation ended with World War II in 1945, but the country was divided along the thirty-eighth parallel, with the northern half occupied by Russia and the south by the United States. When unification negotiations failed, the Republic of Korea was established in the south by free elections in 1948. On June 25, 1950, North Korea invaded the South. The ensuing three-year war, in which U.S. and other U.N. troops as well as China participated, devastated the peninsula.

With the help of vast U.S. economic aid, the ravages of war were repaired by 1960. In 1961, the military under General Chung Hee *Park* seized power. Three five year plans were carried out beginning in 1962. The goal was to rapidly increase living standards by boldly importing foreign capital and technology as well as raw materials for an all-out export drive in manufactured goods, instilling diligence and self-reliance in the people, developing industries, and reinforcing the national defense posture against the threat from the north. This effort has been phenomenally successful.

Foreign Technology and Capital. At the end of the post-Korean War reconstruction, the nation still lacked capital and technological capability for rapid development. The institutional consequences of the developmental hiatus of the thirty-five years of Japanese occupation were still in evidence and not as quickly reparable as the physical infrastructure.

In 1961, Korean exports financed only 15 percent of imports. Yet the dependence of the country on imported raw materials, its high labor productivity, and the relatively low effective internal demand all argued against a policy of import substitution and investment based on technologies appropriate for the domestic market. Korea chose the opposite course. The strategy was:

Phase I: 1962-1971

1. Attract the maximum amount of foreign capital;

2. While using some public foreign loans to provide industrial infrastructure, allow the bulk of the foreign capital to gravitate to export industries, relying on abundant foreign aid to close the balance of payments gap;
3. Enable the labor-intensive industries manufacturing textiles, clothing, plywood, footwear, and electronic parts, enjoying the comparative advantage offered by Korean dexterous and low wage labor, to acquire advanced foreign tools and become highly competitive abroad;

Phase II: 1972-1981

4. Close the balance-of-payments gap by heavy borrowing and by gradually raising exports;
5. Build the domestic technological capability and diversify into heavy and chemical industries, maximizing export earnings by increasing the value added in Korea to imported raw materials and strengthening the independent defense posture of the nation;
6. Develop exports other than to the United States and Japan;
7. Increase domestic incomes and savings to achieve self-financing of development.

Lower labor costs and the availability of highly motivated skilled and unskilled labor were considered as the major asset of Korea by prospective foreign investors, according to a 1972 survey by the Economic Planning Board. Second came the Foreign Capital Inducement Law (FCIL), which offers tax holidays, and guarantees remittance of profits, repatriation of capital, and transfer of royalties.

Other attractions were marketability due to the rapidly expanding internal purchasing power combined with free export zones and government export promotion; the general environment of preferential treatment for foreign investors; and availability of Korean capital from development banks for joint ventures.

The FCIL was enacted in 1962 and covers foreign direct investment, technological assistance agreements, and loans with a repayment period of over three years. It offers the following incentives:

Tax Exemptions: In proportion to the foreign share of investment, the project is exempt from corporate taxes, income taxes, and property acquisition taxes for five years, and to the extent of 50 percent for another three years. Dividends, profit distributions, and royalties received under technological assistance contracts received by foreign investors are exempt to the same extent.

Import Duties: Capital goods imported under the FCIL are exempt from customs duties and the commodity tax, as is a six-month supply of raw materials.

Degree of Control: With the exception of mining and fishing, foreign ownership may be up to 100 percent, although joint ventures are encouraged. All fields are open to foreign investment, with minor exceptions such as the cigarette and ginseng government monopolies.

Guarantee of Remittances: The FCIL guarantees the remittance overseas of profits, capital, royalties, loan principal and interest, and payments for imported capital goods. Foreign currency allocations are made by the Finance Ministry based on applications submitted sixty days before the end of the respective year. Profits may be also reinvested in the country freely up to the amount of the original investment.

Property Guarantees: Private property enjoys constitutional guarantees, and foreign investors must be treated equally as Koreans. The Requisition Law and the Land Expropriation Law which may be applied in emergency situations provide for due compensation. Foreign savings accounted for more than half of investment during 1962-1971. Of this, foreign investments and loans provided $2.7 billion while foreign aid added $1.3 billion. After 1971, foreign aid decreased to nominal levels.

The private sector responded to the incentives offered by the Foreign Capital Inducement Law, as well as the interest rate and exchange rate structures. During 1959-1962, private borrowings abroad amounted to only $2 million. In 1963, they climbed to $55 million and in 1969 to $623 million. Most of this was suppliers' credit to finance Korean capital goods purchases.

Public long-term external borrowing was one measure used to shield the economy from the shock of the quintupling of oil prices in the mid 1970s. The foreign debt grew from $1.8 billion in long-term obligations outstanding in 1970 to $7.2 billion at the end of 1976.

Foreign equity investments amounted to $883 million in June 1977 or 8.1 percent of the total foreign capital received by the country.

The Japanese investments emphasized hotel construction, textiles, electronics and chemicals in that order, while American investments focused on petroleum refining, automobiles, and electronics. In total, the chemical industry received 19 percent of the foreign equity

Table I-2. Foreign Investment 1962-June 1977 (approval basis)
Arrival basis: $713 MM

Country	*Projects*	*Amount ($ million)*	*Share (percent)*
Japan	699	544	62
United States	107	170	19
West Europe	36	80	9
Others[a]	32	89	10
Total	874	883	100

[a]Panama, Hong Kong, Canada, Bermuda, Iran, Bahamas, and IFC.
Source: Economic Planning Board.

investments, tourism 17 percent, electronics 47 percent, textiles 13 percent, metal and machine 10 percent, with the remaining 27 percent being shared by all others. In more than 76 percent of the projects the foreign partner owned more than half of the equity. The major role in foreign financing was played by credit. During 1959-1975, 36 percent ($2.46 billion) was obtained in public loans, 55 percent ($3.8 billion) in commercial loans, and only 9 percent ($.59 billion) in equity investment.

The Economic Planning Board envisions an average annual inflow of $2.6 billion (in current prices) of foreign loans and investments during 1977-1981.

Foreign and domestic industries also have made considerable use of government-created tax shelters in the form of processing for report under bond and free zones. About one third of reports have been produced in this manner in mid 1970s. The device has been particularly effective in attracting foreign investment in the incipient electronics industry, where 64 percent of its 1975 production by value originated in plants enjoying some degree of foreign investment.

Scientific and Technological Infrastructure. The government of Korea was well aware that its plans for economic development based on industrialization were dependent for their success not only on the availability of capital, but also of modern technology. The incipient industries were in poor position to conduct their own research and development. There was a shortage of highly skilled manpower and laboratories, and in most cases production volumes were insufficient to justify the large investment of funds required for original product development or even the perfecting of adaptive technologies. The government therefore took the task upon itself. During the ten years of the first two five year plans, a number of key institutions were created.

1962: KORSTIC (Korean Science and Technology Information Center);
1966: KIST (Korean Institute of Science and Technology);
1967: MOST (Ministry of Science and Technology);
1970: KAIS (Korean Advanced Institute of Science);
ADD (Agency for Defense Development);
1971: KDI (Korean Development Institute).

Science and technology were for the first time recognized by government policy makers as an integral part of the national development plan. The technological infrastructure set up by the government played an important role in the growth of the manufacturing industries, which reached the impressive rates of per annum growth of 14.8 percent during the first quinquennium, 21.1 percent during the second, and 56.9 percent during the third.

The conceptual policy underlying the technological effort has been eloquently stated by the Minister of Science and Technology, Dr. Hyung Sup *Choi*:

> we took the position that domestic R&D is a prerequisite to the sound selection, efficient digestion, and fruitful adaptation of suitable foreign technologies. This is a widely recognized, but certainly hard-to-accept fact of life for almost any developing country. This then implies that priority lies in the sound and efficient utilization of rather readily available foreign technologies to the maximum while guarding against engaging blindly in such, in some instances, noble but certainly financially unrewarding exercises as reinventing the wheel.[1]

By 1976, 86 percent of investment was already being financed from domestic sources. By 1977, the balance of payments gap was closed, and in October of that year the nation had accumulated foreign exchange reserves of $4 billion. With the inauguration of the fourth five year plan, the focus was shifting rapidly towards building up heavy and chemical industries.

Once again the planners realized that these industries were even more dependent on the availability of sophisticated technology. In mid 1977, construction was underway of a second generation of research institutes, which include:

- Standards Research;
- Shiiping and Marine Research;
- Chemical Research;
- Nuclear Energy Development Corporation;
- Machinery and Metallurgical Research (plus the Taejon Machinery Depot);

- Electronic Research;
- Electric Appliance Experimental Station;
- Energy Management;
- Natural Resources Development.

Scientific & Technological Infrastructure

Personnel Sources. Literacy in Korea is over 90 percent. That and other attractive attributes of the labor force stem in part from a well-developed educational system. Standard education consists of compulsory elementary school (six years), junior high school, which is planned to become compulsory (three years), senior high school (three years), and higher education (normally four years). In 1976, 80 percent of primary school graduates went on to middle schools, and of these, 74 percent advanced to high and vocational schools.

Table I-3. Educational Institutions (1975)

Type	*Schools*	*Classes*	*Students*	*Teachers*
Kindergarten	611	878	32,032	2,153
Primary	6,367	98,684	5,599,074	108,126
Middle	1,967	31,441	2,026,823	46,917
High	673	10,840	648,149	20,415
Technical High	479	8,327	474,868	15,340
Technical College	88	1,529	58,500	2,565
Junior College	10	62	3,787	160
Teachers College	16	206	8,504	791
Colleges and universities	72	1,427	208,986	10,080
Graduate schools	82	1,230	13,870	162

Source: Ministry of Education

The institutions of higher learning have expanded rapidly since 1945. Their number has grown fifteen times to reach 268 in 1975. They have been accused of ivory tower mentality and overemphasis on memorization, particularly by industry. In 1969, the national total of 600 M.S. candidates were scattered in 152 departments of twenty-two schools, and the government has sought to consolidate the smaller schools to ease the shortage of laboratories and modern teachers, and to simplify excessive entrance examinations. In recent years the policy has been to improve the quality of higher education on a school-by-school basis rather than across the board. The most noteworthy achievement in this regard has been KAIS (see below), which is patterned after M.I.T. and Cal Tech. In 1976, KAIS conferred 145 M.S. degrees, and other schools some 450. In 1977, nearly 5,000 students received B.S. degrees. Of these, one-third were in science and two-thirds in engineering.

During 1974-1977, the government gave permission to 212 professors and 1,974 students to study abroad. For 1978, a more ambitious program was planned, involving 180 professors and 550 students. Of these, one hundred professors would receive full government scholarships, and thirty have received invitations from abroad. Only fifty of the students will be awarded full government scholarships.

Domestic facilities will be expanded in middle technical education. Some seventeen new junior technical colleges and forty-seven technical high schools and vocational training institutes have been planned, to produce 49,000 technicians and 31,000 craftsmen during the fourth five-year plan (1977-1981). Another 561,000 craftsmen will receive their vocational training in industries over the same period.

In 1975 there were two national libraries, 110 public, 96 special, 136 college and university, and 3,954 school libraries. Among them, they stored twenty million volumes and could seat 350,000 readers. They were staffed by 2,372 professional librarians and 6,018 assistants. They had 9.2 million clients during the year. In addition, there are some 35,000 village minilibraries offering agricultural how-to literature. Bibliographies and catalogues are published by the Central National Library; periodicals index and a list of masters' and doctors' theses are published by the National Assembly Library.

Technology Sources for Individual Firms. The importation of foreign capital equipment has been the most important source of technology for Korean businesses. Machinery and transport equipment represented more than one-fourth of annual imports during 1971-1976, and reached the level of $2.4 billion in 1976. Up until 1977, when for the first time the balance of trade and services was favorable to Korea, the equipment imports were very dependent on foreign export-tied public loans (e.g., Eximbank) and private suppliers' credit. This tended to increase the amount of each transaction as vendors prefer to supply the complete package rather than involve themselves in the integration of imported components with local materials and parts. Once the balance of payments pulled out of chronic deficit, the government began to restrict the use of suppliers' credit. In November 1977, with foreign exchange holdings at $4 billion ($2 billion net), supplier credit was restricted to loans of over seven years, more than $3 million face amount, interest rates not exceeding 1.75 percent over LIBOR (London Inter-Bank Offering Rate) and applications were reviewed with an eye on eliminating turnkey projects.

Technology agreements have been another important source of

know-how. Between 1962 and 1976, there had been 690 such agreements (including 32 cancellations and 278 terminations), entailing $87,574,000 in royalty payments, as follows.

Table I-4. Technology Agreements

Country	*Number of Agreements*	*Percent*	*Royalties Paid*	*Percent*
Japan	460	66.7	$51,693,700	59.0
United States	146	21.1	24,322,900	27.7
West Germany	24	3.5	4,412,300	5.1
Others	60	8.7	7,145,600	8.2

Source: Economic Planning Board, 1977.

In comparison, 865 firms have received foreign investments totaling $954 million. In most cases, but particularly in the newer, technology-intensive industries such as electronics and petrochemicals, substantial technology is involved and financial control or joint venture agreements are used to protect proprietary technology. Between 1965 and 1976 $136 million was remitted abroad as dividends and $25 million as capital repatriation.

Technology Diffusion Networks. Recognizing the importance of technology to Korea as a developing country, both industry and government have made considerable efforts to facilitate the diffusion of acquired technology. The policies summarized above have stimulated the establishment and vigorous functioning of a number of private and public institutions serving as technology diffusion channels.

Industry Associations. To promote their interests, industries have formed a number of voluntary associations. In many cases, technology diffusion has become an added objective pursued with government encouragement. The most important among them play a substantial role in technology transfer, and include:

Federation of Korea Industries. The 1977 membership of FKI consisted of sixty-five associations formed by industries and 283 large corporations. Its Investment Service Center and fourteen binational economic cooperation committees which it has spawned facilitate foreign direct or joint-venture investment. Its Center for Economic and Technical Research carries out studies of long-range international economic cooperation, and performs contract research and joint research with foreign research institutes.

With the help of a UNDP grant, FKI has operated since 1978 an International Management Institute to raise the innovative capability of top and middle managers, and makes arrangements for their additional training overseas.

Korean Traders Association, composed of all government-licensed exporters and importers, has some 2,000 members. Its Trade Information Center offers all types of computerized foreign trade information. KTA also promotes trade fairs and missions, sponsors meetings, research, and publications, and subsidizes the use of foreign consultants.

Korea Federation of Small Business, with a 1975 membership of 23,304 businesses with fewer than 300 employees (half of them under fifty) and assets of less than $1 million each, channels managerial and technical assistance to sixty member firms. A total of 17,586 persons have passed through its training center between 1967-1973. It receives government subsidies, such as a $2 million grant to promote quality control among small industries in 1978 (in cooperation with the KSA).

Korea Management Association specializes in industrial training and management consulting, and publishes a magazine on management.

UNIDO—Korea Association expedites foreign technology transfer to industries by the United Nations Industrial Development Organization.

Korea Marketing Association sponsors research in domestic and international marketing and has 102 business members.

Korea Productivity Center offers services in management and engineering consultation, DEP for MIS, surveys, training aids, publications, exchange of executives with foreign countries under Asia Productivity Organization projects.

Korea Standards Association, with 578 member companies in 1977, acts as a link between government and industry in standardization and quality control programs (publications, training, technical assistance), and is a member of the International Organization for Standardization.

Government-Financed Nonprofit Institutes. The government plan to establish a number of autonomous institutes to facilitate foreign technology reception and diffusion in the country was carried out in the 1960s. It received substantial United States financial aid and technical cooperation (on the order of $10 million per institute). These organizations have attained an unusually high quality of performance and have been successful in their mission. Another characteristic is a high degree of independence from the government, derived from self-perpetuating boards of directors, from special laws which provide for outside rather than governmental audit, and from endowments. They are

Korea Scientific and Technological Information Center (est. 1962). KORSTIC offers a full range of information services, including current listings, reprography, technical consultation, and literature search; it participates in international exchange.

Korea Institute of Science and Technology (est. 1966). KIST conducts research in applied science and industrial technology, techno-economic surveys, and disseminates accumulated technological knowledge for public use. Its contract research for industry covered 1,600 projects by 1976, and accounted for $7.6 million or 60 percent of its income in 1977. Half of the staff of 1,000 are professional scientists. Work in forty-seven laboratories is organized in twelve research groups, and includes:

- Polymer engineering;
- Chemical engineering;
- Organic chemistry;
- Applied chemistry;
- Food and biotechnology;
- Pilot plant;
- Environmental engineering;
- Electrical Engineering;
- Materials science;
- Mechanical engineering;
- Metallurgy and metallurgical engineering;
- Industrial economics.

To staff its program, KIST has carried out a successful reverse brain-drain operation. It maintains close contact with the Battelle Memorial Institute, which was instrumental in the establishment of KIST, and with other scientific institutes overseas.

KIST maintains a number of specialized centers for:

- Software development;
- Technology transfer;
- Foundry technology;
- Precision machinery;
- Electronic communication research.

The services of its Technology Transfer Center include the following:

- Consultation, guidance and arrangements for technology transfer;
- Review of relevance of technology to be transferred (includes evaluation of license agreements prior to government approval);
- Acting as agent for technology transfer;
- Promotion of group technology transfer for medium industries;
- Analysis, evaluation, accumulation, and dissemination of technical information for appropriate technology transfer (includes one-stop consulting service for members);
- Assimilation and improvement of imported technology;
- Follow-up services for imported technology.

Korea Advanced Institute of Science (est. 1971). KAIS is a graduate school of applied science and technology. Annually it admits up to 300 topnotch students for M.S. studies, of whom about 10 percent go on to enroll in Ph.D. studies. Most of its fifty full-time faculty members are Koreans who have received Ph.D. degrees in the United States. KAIS emphasizes extensive interaction with the technological activities of Korean industry. Graduate degree programs are offered in nine fields, which are

- Applied chemistry;
- Applied mathematics;
- Applied physics;
- Biological science and engineering;
- Chemical engineering;
- Electrical engineering;
- Industrial engineering;
- Materials science;
- Mechanical engineering.

KAIS maintains close contact with Stanford Research Institute, and invites foreign educators to serve as visiting professors.

Korea Development Institute (est. 1971). KDI conducts research on policy matters concerning the national economy, helps develop the Korean five year plans, and advises economic policy makers. Sixteen full-time senior fellows holding foreign Ph.D. degrees are in charge of research projects. It conducts studies in collaboration with foreign research centers, such as "Korean Modernization: 1945-1975," undertaken with the Harvard Institute for International Development. It also exchanges research results with more than forty foreign institutes, holds international symposia and hosts visiting economists. Its 30,000 volume library has exchange arrangements with sixty foreign and thirty domestic economic institutions.

Korea Design and Packaging Center (est. 1970) conducts research in design and packaging, including collecting up-to-date information abroad, and trains designers and package engineers and organizes exhibitions primarily for export industries.

Korea Trade Promotion Corporation (est. 1962) provides trade information services to Korean and foreign businessmen, facilitates business transactions and marketing, and organizes trade fairs. KOTRA has overseas branches in seventy-six cities.

Political Climate for Business

Local Political Climate for Business. Under the 1972 Yushin constitution, Korea has a presidential system of government under which the president controls the executive branch, dominates the legislature and takes certain crucial powers away from the judiciary. President Chung Hee *Park* has been in office since 1961. His staying power is protected by reason of electing the president every six years by the national Conference for Unification rather than by direct vote. The conference has 2,000 to 5,000 delegates elected by popular vote every six years. In addition to electing the president, the conference elects one-third of the unicameral six year national assembly on the recommendation of the president for three-year terms. The president also appoints the nine members of the constitution committee, which passes judgment on the constitutionality of laws, matters of impeachment and dissolution of political parties. Opposition leaders, some of whom have been in jail, having committed the unpardonable sin of losing out in the political contest, could no doubt bear witness to the shortcomings of this system with regard to human rights. There is also much evidence, however, that the stability that the Korean system of government has provided has been an important element in bolstering business confidence and economic growth. Power to rule by emergency decree has helped face crises such as the oil price shock of 1973.

From this power position since 1962, the Korean government has provided the private sector with stable rules of the game, increasingly adept economic planning, effective monetary policies, and a system of tax incentives, support institutions, and public investment in physical and social infrastructure, rather than controls and penalties.

Relations with North Korea, United States and Japan. Koreans do not consider themselves free of factionalism nor as a people prone to compromise. Thus, it has been a blessing in disguise that in its effort to focus the attention of the nation on productive work rather than internal squabbling, the government has been aided mightily by the external threat posed by the North Korean Communist regime. National unity has been easier to maintain in the face of potential outside aggression than it would have been otherwise. *Saemaul Undong* (new community movement) has been an important motivator of rural development (just as Israeli *kubbutzim* thrived under Arab guns), and in recent years is being extended into urban areas. This not only produces civic campaigns such as park cleanup drives but also promotes labor peace in factories. The economic bankruptcy of North Korea offers an additional argument. When President Park in his 1977 New Year's speech offered food donations to North Korea, the implications were hardly lost on his South Korean listeners.

Since the Korean conflict, relations with the United States have been close and cordial. American aid was essential to Korean postwar economic recovery, and its scale was generous. Because of the presence of a large U.S. military establishment for thirty-two years, a whole generation of Koreans has had intimate exposure to American culture, values, and technology. Familiarity with the English language, albeit not profound, has facilitated technology transfer.

In 1977, relations came under a passing cloud when the planned gradual withdrawal of U.S. troops was announced. They were also muddied by the Tong Sun *Park* scandal, which implicated the Korean government in allegations of influence buying in Washington. Most Koreans, however, take the longer view, remembering the blood kinship formed on the field of battle and being aware of the strength of mutual interest. One of the reasons why Korean business leaders are working to attract American investment is to see their country more safely under the American military umbrella along with an increased American stake.

With regard to Japan, Korean feelings tend to be a mixture of love and hate. The hate stems from the occupation period between the two World Wars, silhouetted against the historical backdrop of previous Japanese invasions. The love grows out of admiration for Japanese accomplishments since 1945, fed by affection deriving from long familiarity with the Japanese language, culture, and customs. Koreans are fond of comparing the structure of their economy and society with the Japan of ten years ago. Much of the technology that is acquired by Korea arrives via Japan—a route preferred by most businessmen both because of the adaptation that it undergoes in Japan as well as the ease of communication. (The government and Korean exporters think, however, that such Japanese influence is a mixed blessing, because of the time lag involved and because of Japanese competitive discrimination against Korean exporters to Japan.)

Economic Factors

In 1962, Korea was among the poorest of nations, dependent on its primitive agriculture to produce 40 percent of the national income and on foreign grants to close a chronic gap in its balance of payments. By 1977, it was a semi-industrialized, middle income nation. The dimensions of the change are illustrated by the data in Table I-5.

To understand how this stunning achievement has come about, the data in Table I-6 will be helpful. Exports, which amounted to a mere $163 million in 1962, exceeded $10 billion in 1977, rising from 2 to 36 percent of GNP.

An interesting feature of this strategy has been that it made the usual dilemma between growth and equity irrelevant. Driven by the forces of comparative advantage, Korea has been expanding primarily in labor intensive manufacturing activities, relying on the elasticity of the export market to absorb 23 percent of total manufacturing production by 1975. During the past fifteen years, the capital output ratio has remained practically constant at 1.3; and so has the cost of job creation at $5,000. As a result, the benefits of growth have been distributed widely. Real wages in manufacturing have been increasing by an average of 8 percent during 1963-1976. At the same time, government price subsidy policies increased farmer purchasing power (2 percent per annum over parity during the fifteen years) to supplement the annual growth of real rural income of 6 percent.

Table I-5. Economic Factors of Korea
(in 1975 prices)

	1962		*1976*
GNP ($ billion)	5.5		21.6
GNP real annual growth rate (percent)		9.7	
Share of manufacturing in GNP (percent)	14.5		30.0
Gross investment as percent of GNP	13		28
Domestic savings as percent of gross investment	33		86
Domestic savings as percent of GNP	2		20
Unemployment	(1963) 8.2		3.9

Source: Economic Planning Board.

Table I-6. Average Annual Real Growth Rates 1963-1976
(in 1975 prices)

Exports: total	34.6
Manufacturers	45.0
Employment	3.9
Productivity: value added: total	10.3
Manufacturing	18.6
Value added per worker: total	6.4
Manufacturing	6.4
Gross investment	17.0

The Korean economic miracle cannot be explained without taking into account the independent variables of political will, social discipline, and unusual availability of large inflows of foreign capital. Between 1966 and 1976, long-term foreign debt outstanding rose from $300 million to $7 billion, apart from earlier grant aid of some $12 billion.

It is necessary to acknowledge also that the Korean economic strategy has worked exceedingly well. It is a tribute to the Korean economic planners led by the Minister of Economic Planning Vice Premier Duck Woo *Nam* that they recognized from the outset that Korea's economic future rested on two foundations: export-oriented growth as the best available method of overcoming the twin handicaps of a small internal market and lack of domestic natural resources; and labor-intensive manufacturing as the source of Korea's comparative advantage.

The export drive has been nurtured by suitable monetary policies that maintained realistic exchange rates of the *won* by drastic devaluations, and provided foreign currency loans to exporters for the import of raw materials at concessional rates and offering special facilities to small business. At the same time, inflation has been kept within bounds by high interest rates, which attracted savings, thus reducing effective demand, and by public utility pricing at full cost levels.

Table I-7. Korean Economy: Average Annual Rate of Change (percent)

	1962-66	*1967-71*	*1972-75*	*1976*
Wholesale price index	16.3	7.8	22.4	12.1
Seoul consumer price index	16.2	11.4	16.2	15.4

Physical Infrastructure

Energy. Given the rapidly expanding economy, demand for energy has been increasing also (by 7.9 percent per annum). It rose 2.7 times from 20.2 in 1962 to 54.4 in 1975 in terms of million metric tons of coal equivalent. Increased industrial use has also shifted the consumption patterns away from the traditional firewood and charcoal to coal and oil.

The share of oil has increased from 9.5 percent in 1962 to 52.7 percent in 1975. Domestic production of anthracite coal, spurred by the rising oil prices beginning in 1973, increased 2.4 times between 1962 and 1975, to 17.6 million metric tons, but still accounts for only 33.2 percent of total energy consumption. Hydroelectric power and atomic energy provided less than 2 percent, but the latter will grow as five plants begin operations between 1977-1984. Electric power generation capacity increased 10.9 times to 4.7 million kilowatt/hour during the 1962-1975 period and has stayed ahead of demand.

Transportation. Korea has 10,000 kilometers of paved roads and 35,000 kilometers of other roads. Six modern expressways link all regions into a one day travel network of 1,142 kilometers. In 1975, 200,000 vehicles were in use. There are 5,619 kilometers of railroads. Since the war, the rolling stock has been replenished and modernized and includes 375 diesel locomotives, 66 electric engines, and 128 electric motor coaches.

Marine transport has of course played an important role in the export trade. At the end of 1975, 17,211 ships were registered, with a capacity of two million gross tons. The main ports are Busan in the south and Inchon near Seoul.

Korean National Airlines provides air transport internally as well as internationally. KAL provides cargo and passenger service to four Japanese cities, Hong Kong, Manila, Taipei, Singapore, Paris (polar route), Honolulu and Los Angeles. In 1975 it had twenty-five airplanes including three B-747s. In 1975 railways carried 52 percent of domestic freight (in ton per kilometer) and 25 percent of passengers (in passenger per kilometer); roads carried 22 percent of freight and 73 percent of passengers, ships carried 26 percent of freight and 1 percent of passengers, while airplanes carried 1 percent of passengers.

Communications. In 1975 Korea had 1.2 million telephone circuits, 2,000 post offices with daily delivery, a microwave intercity telex system, and semiautomatic telephone exchanges with Japan and the United States. The telegraph system handled thirty-four million domestic and one million international messages.

Cultural Patterns of Business Activity

Social Characteristics. Korean society may be characterized by an acceptance of differences among individuals. The Confucian acceptance of social hierarchy has made it easier to maintain labor discipline as well as political quiescence. At the same time, there are no strong regional, ethnic, caste, or religious cleavages in Korean society. The Korean labor force is mobile, adaptable, and broadly educated.

The Japanese occupation of 1910-1945 has had a leveling social effect. While new private fortunes have been made during the economic boom of the 1960s and 1970s, there is no old money and privilege. As a consequence, income distribution compares favorably with such countries as Great Britain and Sweden. A 1970 International Bank for Reconstruction and Development survey showed that the lowest 40 percent of income earners accounted for 18 percent of total income, while the top 20 percent received 45 percent.

Worker Attitudes. Younger Korean workers are mobile and respond quickly to monetary incentives. Once married, however, the customary inclination is to stay put. They look to their employers to provide job security. Rapid growth of industries, while creating new job openings on the lower levels, also has speeded up the process of in-company promotion and has contributed to employee permanence. Average labor turnover in industry has been only 4 to 5 percent per annum.

Wage rates are but a fraction of the standards of the industrialized countries. Estimates put them at about one-fifth of Japanese and one-tenth of American wages. Earnings are increased by working hours that at fifty-two hours per week and 25.2 days per month (all industries in 1976) rate among the longest in the world.

Internal Migration. Industrialization has changed the distribution of the population in Korea. While in 1949 only 17.2 percent lived in cities, urban population in 1975 represented 48.4 percent of the total, (which grew from twenty to thirty-five million). There are thirty-five cities with population exceeding 50,000, but more than 63 percent of the urban population live in the principal cities of Seoul, Busan, and Taegu.

In the 1970s, a successful effort has been made to slow down the migration to Seoul by providing new industrial parks in other areas of the country, and by raising farm incomes.

Dominant Industries and Markets

Traditional Industries. In the 1960s Korean industries responded to the price elasticity of external demand. Due to low wage scales, Korean labor intensive industries enjoyed a notable price advantage competing in export markets. As a result, considerable growth ensued in established low technology light industries—textiles, clothing, footwear, and plywood. Around 1970 annual production growth rates were being registered as follows: 13 percent for cotton yarn, 10 percent for cotton fabrics, 103 percent for garments, and 30 percent for plywood. In addition, production in the new electronics industry was growing by $56 per annum.

Markets. Export markets were absorbing large portions of the production of these industries. By 1976 the five light industries accounted for more than half of all exports. All industrial products then constituted 88 percent of exports, as compared with 27 percent in 1962.

By 1976, the traditional exports were running into difficulties in the export markets. The Japanese market had been much less open to manufacturers from the beginning, as evidenced by the fact that its share of Korean exports fell from 43 percent in 1962 to 23 percent in 1976. The EEC and the United States also began imposing stricter tariff and quota restrictions to stem the flood of low-priced Korean products, which were causing dislocations in their own industries. Plywood exports were especially hard hit by the worldwide housing slump which followed the 1973 oil crisis.

Table I-8. Major Korean Export Commodities (1976)

Category	*Value ($ Million)*	*Percent of Total Exports*
Textile products	2,735	35
Electronics	801	10
Shoes	398	5
Plywood and wood products	361	5
Iron and steel products	354	5
Fish and shellfish	318	4
Transport equipment	248	3
All others	2,151	33

New Industries. Korean planners had anticipated the market resistances to some levels of traditional exports, and realized also that increasing domestic political pressure for distribution of the increased income would gradually blunt the price advantage of labor-intensive exports.

Beginning with the third five year plan (1972-1976), new priority industries were selected: iron and steel, transport equipment, petrochemicals, machinery, as well as the already advancing electronics. The concept is to attain further growth of exports (from $10 billion in 1977 to $20 billion in 1981) by gradually substituting high technology for cheap labor as the comparative advantage in exports. Toward this end, $1 billion has been invested in the government-owned Pohang Steel Company which was producing 2.6 million tons of crude steel by mid 1976, doubling Korean capacity. Built with imported equipment and know-how, it is considered to be one of the most modern and well-designed steel mills in the world. Shipbuilding capacity has risen from 189,000 gross tons in 1971 to 2,390,000 gross tons in 1975. The automobile industry recently produced the first Korean-engineered sedan.

The Ulsan petrochemical complex with its 100,000 ton naphtha-cracking center and seventeen downstream product plants was inaugurated in 1972. Another complex at Yochon with ten production plants centered on a 350,000 ton naphtha cracker is scheduled to be completed in 1979.

By mid 1976 there were 2,104 machine-producing establishments, and since 1970, industry production has doubled every year on the average.

New Markets. The changing structure of Korean production has called attention to new markets. This trend has been further promoted by planners who saw increased risks in the rising dependence on foreign trade, and welcomed diversification of export markets. Between 1962 and 1976, the Japanese share of Korean exports declined from 43 to 23 percent while the U.S. share grew from 22 to 33 percent, and all others rose from 35 to 44 percent.

Between 1975 and 1977, significant breakthroughs were made in the Middle East, mainly by exporting construction services which included Korean labor and cement. Contract awards reached the level of $2.6 billion per annum in 1976.

For the future, Korea expects to follow in Japanese footsteps as an adaptor of Western technology and a technological intermediary between the West and the developing countries.

Issues Affecting Firms

Business Relations with Government. In addition to the external threat posed by North Korea, Koreans respond to the national opportunities that peace time offers. Having been buffetted by numerous aggressions since the Hideyoshi invasion in 1592, they are quick to seize the historical intervals between harsh times to build, produce, and trade. This promotes a kind of patriotic alliance between business and government, and in the 1960s and 1970s, the Korean government and business have been very supportive of each other's goals. Most companies have integrated the guidelines of the national five year plans into their own plans, and attach great prestige to such symbols as the Prime Minister's citation for exceptional export performance. The government in turn keeps its big stick in the closet and uses the carrot instead, in the form of tax, credit, and procurement award incentives.

Attitude Toward Foreign Technology. It is only very recently that the Korean government began to make a sustained effort to introduce Western science through the higher levels of the educational system. It will be a long time before Korea can be expected to have an independent capability to create new technology across the board. In the meantime, foreign technology is being introduced at the plant level on a substantial scale. Most managers, however, lack technical training, and their engineers do not always understand the scientific principles behind the technology that is being imported.

As a consequence, the tendency of both top and middle managers is to absorb the new know-how undigested. The process could be compared with inserting a cartridge into a tape deck. To assure proper functioning, great care is taken to replicate the prescribed process without modification. However, the capability to adapt the formula if it does not work properly is missing because of differences in raw materials, skills, and other environmental factors. The Ministry of Science and Technology publishes annual lists of technologies that firms are encouraged to acquire from abroad. This list includes the addresses of foreign owners and in 1977, cited 1,758 items for acquisition.

Size. Some 30 percent of Korean exports are produced by small enterprises that, as noted earlier, are defined as having fewer than 300 employees or assets of less than $1 million. The goods are increasingly complex manufactures, and the inevitable problem is quality control, which the smaller businesses are ill-equipped to provide. Large domestic assemblers of subcontracted components (e.g., in the automobile industry) encounter similar problems.

The government is attempting to help rationalize small business by providing industrial parks, credit for new equipment, and designating only a few firms to produce an identical item. The Federation of Small Business is promoting cooperative efforts and better management, and the KSA provides training in quality control and assistance in standardization. Problems, however, stem from the fact that many of the small businessmen enjoy their social status as independent bosses and do not wish to become mere executives of larger companies by merger or absorption. They cannot afford to hire topnotch managers who command high salaries in big industries, nor does the scale of their operations warrant the expenditure that would be necessary to plan systematically. While in the long run rationalization of the small firm by consolidations may be inevitable, to date this seems to have occurred more in the form of conglomerate groups owned by wealthy families and sharing only financial management than by true fusion by product line to achieve economies of scale. Owner families appear to be conservative in their thinking and more interested in protecting the moderate average profitability of their investments by diversification than in maximizing profits by horizontal integration.

Labor. Korean labor unions date their history from 1953. Only some 10 percent of the urban labor force has been unionized, mainly in large manufacturing enterprises. The Korean Labor Standard Law establishes working conditions in enterprises with more than fifteen employees. It stipulates a labor contract with each employee covering wages, hours, and conditions; thirty day dismissal notice, and thirty days' severance pay for each year worked; payment of 150 percent of wages for overtime work, and 60 percent during periods of voluntary shutdown; forty-eight to sixty hour week (excluding rest hour); fourteen legal holidays and paid vacations based on seniority and attendance record; preventive health and safety measures including annual physical examinations; workmen's compensation insurance; and paid maternity leave and limited hours for women and minors.

The Labor Disputes Adjustment Law provides for mediation and arbitration of disputes during a twenty to thirty day cooling off period during which strikes and lockouts are prohibited.

There is no legal minimum wage. In June 1977, average monthly earnings in industry were $184, where the work month was 25.4 days. Average monthly earnings in 1976 were

in manufacturing	$107
in transport	133
in mining	136
in commerce	155
in social and personal services	203
in construction	238
in finance, insurance and business	254
in electricity, gas and water	275.

BRAZIL

Economic Profile of Brazil

Introduction. Brazil, with a population of over 110 million people, is South America's largest country, in both population and territory. The growth in population has enlarged the domestic market for a variety of local and foreign consumer and industrial goods. Yet several factors, including small scales of operation, have resulted in higher unit costs and higher prices to the user of Brazilian produced goods.

Brazil has some 3,000 miles of beaches on the Atlantic coastline, which, coupled with its warm climate, make the country an almost matchless tourist attraction. The recreational and tourist industry can provide significant foreign exchange earnings. The country also has vast mineral resources that are barely tapped—the Amazon River Basin alone contains the world's largest known reserves of iron ore; promising deposits of oil have been discovered near Rio de Janeiro and at the mouth of the Amazon River. Clearly, foreign capital and technology will be called on to aid in the development of these vast natural resources.

Brazil's land mass is greater than that of the United States. It also has more land under cultivation than all European countries combined. The nation is the world's largest producer of coffee, bananas, and sugar; second largest producer of soy beans; and third largest producer of cocoa and tobacco. In the agricultural area there is a great deal of interest on the part of the Brazilian government, trade associations, and institutes to industrialize more fully the operations of national producing units. Specifically, there is interest in working towards a greater mechanization of the growing, harvesting, transportation, and refining processes (in most cases involving the utilization of capital-intensive technology) and towards the development of more industrialized agricultural products.

The Brazilian government has made it clear that manufactured

Table I-9. Economic Profile of Brazil

(1976 Official Statistics–Source: *Brazilian Business*)	
GNP growth (percent)	8.8
Industry (percent)	11.0
Agriculture (percent)	4.2
Balance of payments	
Trade balance (U.S. $ billion)	–2.2
Imports (U.S. $ billion)	12.3
Exports (U.S. $ billion)	10.1
Current transactions (U.S. $ billion)	–6.0
International reserves (U.S. $ billion)	6.47
Direct foreign investments (U.S. $ billion)	1.0
Inflation	
General price index (percent)	46.3
Cost of living–Rio (percent)	44.8
Cost of living–Sao Paulo (percent)	37.9
Wholesale Price Index (percent)	44.9
Construction costs–Rio (percent)	58.6
Means of payment expansion (percent)	36.4
Credit expansion to private sector (percent)	55.8
BNDE financings–disbursements (percent)	45.8
National Treasury cash surplus (Cr$ million)	429.0

products will play an important role in the overall development of the economy and in improving the national balance of payments. Although the government is stressing the development of an industrial economy, the largest share of total national production (almost 70 percent) is now realized through the sale of manufactured exports. The top ten of these products in Brazil in 1976 were

- Transportation equipment,
- Heavy machinery and equipment,
- Processed coffee,
- Electrical equipment,
- Shoes,
- Processed meats,
- Refined sugar,
- Orange juice,
- Cotton yarn, and
- Office equipment.

Brazil exports a variety of goods, but in only a few instances has the nation developed significant markets in other countries. Exports for 1977 rose to $12,139 million.

The Brazilian government views the national firm as critical to the economic development of the nation. The government has taken a number of initiatives to make sufficient funds available to allow capital-scarce national firms to either start up or expand operations in government established priority areas. The problems of technology development and acquisition are more difficult to resolve. While various national, state, and university groups are conducting, or plan to conduct, programs designed to strengthen the administrative and technical capacity of the Brazilian national firm, these and other technology initiatives have so far had only a modest impact on the operations of the national enterprise. Limited educational and training facilities, a tendency to adhere to traditional ways of operating business, and lack of research and development resources (equipment, competent technical staffs, and capital) are cited as major reasons for this situation.

Balance of Payments and Inflation Position of Brazil. Until 1973, the Brazilian economy performed strongly. The real growth rate of the national product was above 10 percent, and the annual inflation rate was approximately 20 percent.[2] This was a dramatic change from 1964, when the inflation rate was almost 100 percent, and the economic growth rate barely approached 5 percent.[3]

During this pre-1973 period, Brazil's balance of payments position was quite strong. In 1965, international transactions accounted for 10 percent of the national product, while in 1973, they accounted for 17 percent. This trade performance strengthened Brazil's foreign reserve position, with the balance in this account standing at U.S. $4.2 billion at the beginning of 1973.[4]

However, the oil crisis of 1973-1974 presented major problems for Brazil, and Brazilian imports rose dramatically in 1974. During that year imports increased more than 104 percent while exports advanced only 28 percent.[5] The dramatic rise in imports resulted in a current account deficit of U.S. $4.69 billion and in a transaction account deficit of U.S. $7.122 billion.[6]

Since the end of 1973, Brazil has made a concerted effort to bring its balance of payments position and inflation back to acceptable levels. To achieve this, the government has raised customs duties, forbidden imports of nonessential items, required a 360-day deposit on imported goods, implemented various fiscal and credit programs to stimulate exports, and allocated billions of dollars to investment projects in an attempt to produce locally what the country has bought abroad in the past. Although these measures have reduced inflation somewhat, the inflation rate for 1977 was 38 percent and is in 1978 about 40 percent.

Brazil's credit rating abroad remains stable largely because of confidence inspired among creditors by the government's tighter monetary and fiscal controls. As a result, despite Brazil's foreign debt of $32 billion (early 1978), there should be few, if any obstacles to raise the funds required for development activities.

Energy Performance. Energy remains a critical problem for Brazil. Brazil imported 80 percent of its petroleum in 1978.

The task of meeting the energy needs of Brazil is primarily shouldered by the large government agencies such as Petrobras and Nuclebras. The Minister of Energy and Mines, Shigeaki Ueki, has been a strong proponent of developing nonoil sources of energy, particularly nuclear and hydroelectric sources.

Investment Programs of Government. The total investment programs of the Ministries of Industry and Commerce, Interior, Social Welfare, Mines and Energy, Communications, and Transportation were reduced by 25 percent in 1977, based on the government's review of the country's balance of payments position, the rate of inflation, and the oil crisis.

Under the guidelines of the economic plan for 1977, priority investments and permission for importations were extended to the following industry groups:

- Petroleum,
- Capital goods,
- Steel,
- Nonferrous metals
- Chemicals and petrochemicals,
- Fertilizer production,
- Paper and cellulose, and
- Agricultural pesticides industries.

Applications for funds for these investments are overseen by the Economic Development Bank (BNDE) with priority being given to the development of the capital goods industries. Funding of small and medium-size Brazilian national firms operating in priority areas is given particular attention by BNDE.

Other Environmental Factors. It is estimated that fully 40 percent of Brazil's population is illiterate. Although increasing numbers of Brazilians are attending educational institutions, the quality of the programs offered has been declining.[7]

There is growing concern in the country about the nation's

unemployment rate. Slow growth in the economy and a rapidly expanding population base have been cited as the major causes of the nation's high unemployment. The government is attempting in some sectors to retain and reemploy displaced workers, but such programs have not met with considerable success.

Brazil has wide variations in income levels. This is particularly evident in cities such as Rio de Janeiro where expensive condominiums are situated within a few miles of slums (*favelas*).

Brazil's Development Orientation, Policies, and Strategies

Introduction. Some have described the Brazilian system as being relatively relaxed in its control over both internal markets and direct foreign investment and foreign technology. Others have concluded that especially in the area of technology transfers Brazil is becoming less liberal even though science and technology have become since the enactment of the National Development Plan of 1972 topics of major governmental concern. Whichever of these observations is correct, Brazil is a country with vast potential, a country that needs to acquire more modern technology, a country whose government is keenly interested in economic growth and development, and a country where international companies will continue to hold major investments.

The national goals of the Brazilian government, as stated in the National Development Plan II (1975-1979), have been translated into specific development strategies. Among these are the following:[8]

- Construction of a strong industrial base, through a policy of mergers and incorporations and formation of industrial and financial conglomerates to strengthen certain key sectors;
- Absorption of foreign capital, technology, and managerial capacity to complement the domestic efforts to generate these factors of production;
- Modern industrial technology to increase the competitiveness of industrial sectors;
- Creation and adaptation of technology;
- Strengthening the national firm's capacity to compete with foreign enterprises;
- Modernization and reorganization of certain traditional industries;
- Development of basic sectors, particularly in capital goods industries, basic electronics, and basic inputs;
- Opening of new markets for manufactured exports.

In the coming years, continued economic expansion by Brazilian industry will depend in great measure on how effectively energy problems are resolved and the degree to which Brazil can acquire capital and technological assistance from abroad.

Brazilian Government Policy Toward Foreign Investment. Brazilian government policy towards foreign investment has varied considerably over the past fifty years. During the period from the mid 1920s until the end of the Getulio Vargas regime in the 1940s, the government strategy was generally that of exclusion of foreign interests. From the mid 1940s until the military's assumption of power in 1964, Brazil moved to regulate foreign investment. Such regulation took the form of strict controls on profit remittances to foreign owners. It also favored public sector financing of firms with a majority Brazilian ownership and prohibited governmental loan guarantees for companies with majority foreign equity.

Since 1964, the Brazilian government has followed a program of providing incentives for increased foreign investment in certain industrial sectors and geographical regions. It has also worked hard to increase Brazilian participation in certain key economic sectors. Brazil today welcomes foreign investment.

Brazilian Government Position Toward Technology.[9] The Second Basic Plan for Scientific and Technological Development (PBDCT 1975-1977) estimates that approximately $1.5 billion were spent on government technology programs by the end of 1977. Table I-10 indicates the manner in which such government funds have been allocated.[10]

In an effort to increase the flow of technology transfer from

Table I-10. Brazilian Priority Programs

Priority Programs	*Percent of Total Technology Investment*
Industrial Technology	28.0
Basic Research and Postgraduate Studies	22.0
Development of New Technologies	16.4
Agricultural Technology	11.1
Infrastructure Technology (energy, transportation, communication)	10.0
Support Activities	4.9
Special Integrated Projects and Applied Technology for Social Development	4.7
Planning and Programs Under Study	2.9
	100.0

foreign sources and to spur economic development, Brazil is allocating millions of cruzeiros annually both to acquire know-how and to evaluate specific technology programs and transfer mechanisms. For example, during 1974, the government spent approximately 3.1 billion cruzeiros (1975 value) on the development of advanced technology programs for the nation.[11]

Brazil's specific objectives in the technology area[12] are based on three major considerations.

It is necessary to keep a balance between fundamental research, applied research and development.

The bulk of the effort must be directed to updating the technology in a number of different sectors, bringing to them the benefit of knowledge already existing in developed countries and carrying out greater or lesser adaptations of technology.

Scientific and technological policies must at the same time concern themselves with avoiding waste of natural resources.

Priority areas for new technology development include nuclear energy, nonconventional sources of energy, space activities, and oceanography. Continued technological development is promoted in the basic industrial sectors such as the electronics, capital goods, chemical and petrochemicals, steel and metallurgical, and aeronautical industries. (See Table I-11 for other priority areas.)

Direct Brazilian Government Programs to Aid Technological Development

Governmental support in the technology area can be grouped into four categories[13] as follows.

Direct Financial Assistance to Producers and Users of Technology for the Selection, Research, and Development of Specific Products or Processes. This type of assistance involves three federal organizations: FINEP, STI (Secretariat of Industrial Technology), and BNDE (FUNTEC). The financial assistance provided by FINEP has been directed largely toward priority areas within the capital goods industries. Programs funded have involved activities such as new product development, the testing of prototypes, and the construction of pilot installations. FINEP has traditionally funded preinvestment studies which have allowed industry groups to study potential projects before taking final action.

STI (Secretariat of Industrial Technology) is part of the Ministry of Industry and Commerce. Its grants have been principally directed

Table I-11. Major Priority Areas for Technology Development
(partial listing)

1. The development of new techniques in
   - nuclear energy
   - nonconventional sources of energy
   - space activities
   - oceanography

   These generally are areas of large scale, high cost operations.
2. The technological development of basic industrial sectors of high technological content:
   - electronics industry, in particular computers
   - capital goods industry
   - chemical and petrochemical industry
   - steel and metallurgical industry
   - aeronautical industry
3. The development, especially by large scale national and foreign enterprises, of their own research effort.
4. The development of infrastructure technology to maintain the sectors of energy, transport, and communications on an equal level with international technology.
5. The development of agricultural technology to conduct a permanent and intensive program of research on the basic products.
6. The development of technology applied to regional development.
7. The development of technology of social areas with an emphasis on health and education.
8. The guarantee of the high quality of research undertaken.
9. Identification of the areas most in need of basic research.
10. The selection of technologies to be imported according to priority criteria.
11. The flexible use of the world system of patents.
12. The creation of the stimulus to adapt, absorb, and diffuse imported technology through effective means and procedures.
13. The creation of fiscal and/or financial incentives to induce all large enterprises to establish their own research budgets for technological adaptations or innovations.
14. The creation by different sectors of research institutions financed by businessmen themselves.
15. Implementation of a national postgraduation plan integrating graduating graduate studies and research in universities.
16. Expansion of international technical cooperation designed to import brains and advanced technologies and to provide postgraduate training abroad for Brazilian technicians and teachers.

toward public sector institutes in an effort to strengthen research and development capabilities. STI activities have centered on the development of product lines (within such areas such as metallurgy, chemicals, electronics) and basic infrastructure (transportation, energy). FUNTEC's purpose is to assist technological development at the firm level and to stimulate research groups to direct their attention to the needs of the national economy. Its funds are available to a wide range of industry groups and research organizations.

Direct Technical Assistance in the Form of Specific Technical, Consulting and Engineering Services to Enable the Productive Sector to Adapt New and/or Existing Technologies. The government's direct technical assistance programs are designed to provide national groups with services needed for the expansion of technological development.

FINEP, through its program of support for national consulting, has strengthened the capacity of national consulting firms to assist in the development of new techniques. STI recently has extended its support for efforts to strengthen Brazil's national engineering capacity. It is hoped such efforts will lead to a reduction in capital equipment imports.

Scientific and Technological Information Services Designed to Disseminate Data on Current Know-How as Well as to Assist Users and Producers of Technology in Locating Sources and Markets for Technology. The National System for Scientific and Technological Information, under the direction of CNPq, is still in the developmental stage. More resources will have to be committed to information services to enable Brazilian industry and government agencies to obtain current data on technology developments in their fields. INPI currently maintains its own information system concerning registered patents and trademarks.[14]

Direct Research and Development Programs of Governmental and Quasi-Governmental Groups, Principally in High Priority Economic Sectors. The Brazilian government's research and development programs are being counted on to increase indigenous technological capabilities with locally developed know-how. Such programs involve almost every ministry of the Brazilian government and are primarily directed to achievement of national goals and policies. Particular emphasis is now being given to research and development in the fields of agriculture, communications, transportation, and military weapons. Private sector institutions frequently participate in the government programs.

One example of the government-sponsored research and development program is the IPT (Institute of Technological Research). This type of institution, along with universities of Sao Paulo and Campinas, is a major source of locally developed technology.

Indirect Brazilian Government Programs That Have Strengthened the Nation's Technological Base

Indirect governmental support in the technology area can be grouped into six categories:[15]

Programs to Strengthen the Financial Position of the Brazilian Firm. The Brazilian government has undertaken financial programs resulting in the participation of certain semiautonomous agencies in the long-range capital funding of existing firms. EMBRAMEC, IBRASA, and FIBASE (all BNDE organizational groups) assume minority interests in national companies. Their participation is intended to influence the firm's business strategies and overall administration.

Measures to Stimulate Domestic and Foreign Market Demand. The Brazilian government has not extensively utilized procurement as a tool to stimulate technology development. Certain ministries are involved in supplier development programs but governmental consumption is not particularly oriented to the development of new and/or improved technologies by national industry.

Support for nationally produced capital equipment has been traditionally provided by FINAME (a BNDE agency) through the extension of governmental credits. FINAME uses funds supplied by BNDE and other financing agencies for medium, long-term, and special credits in connection with the purchased machinery and equipment.

The development of technology and increased productive efficiency has been stimulated by governmental export incentives. Such incentives, administered by CACEX (of the Banco do Brasil), provide exporters with easy access to credit facilities at interest rates more favorable than those in the domestic market.

Programs to Establish Standards and Norms. INMETRO (National Institute for Metrology, Standardization and Industrial Quality) is responsible for carrying out consumer protection actions and certifying product quality standards. INMETRO is under the Secretariat for Industrial Technology. CONMETRO, the interministerial council for establishing national norms, oversees the activities of INMETRO.

Programs to Modernize and Develop Industrial Operations Through Various Regulatory, Tax, and Incentive Programs. *Mergers* *and Consolidations:* The Brazilian government is attempting to stimulate mergers and consolidations to achieve, among other things, a greater level of economic efficiency and lower unit production costs and to increase the potential for technological development. In those cases in which mergers and consolidations are considered to be in the interests of the national economy, the government will grant to companies the option of revaluing fixed assets above the limits of the monetary correction and up to the market value, without paying any income tax.

Mergers and consolidations that have been granted income tax exemption have resulted in larger scales of operation, lower unit costs, and lower prices; stronger competitive positions because of improved management, increased technical capacity, more effective research and development efforts; development of new foreign markets; installation or modernization of enterprises in identified priority economic areas.

Joint Ventures: The Brazilian government favors the joint venture on grounds that it provides the Brazilian firm with an opportunity to strengthen its management, technical, and/or financial capacity. The trend in Brazil today is toward the formation of joint ventures.

Trading Companies: This form of enterprise is heavily supported by the Brazilian government in its efforts to establish and expand foreign trade. Trading companies are considered important mechanisms for foreign trade expansion because most Brazilian firms do not have the individual capacity to compete effectively in foreign markets. The Brazilian law grants tax incentives to both the producers and the trading company in connection with their foreign trade initiatives. These include tax benefits for the producer on goods for export, profit tax relief, export tax credit for the trading company, and recording of insurance and transportation costs as additions to the value of foreign sales, if these services are provided by Brazilian institutions.

Tax Incentive System. *Tax Incentives.* Companies may invest a stated percentage of their income tax in

- Projects approved by the development agencies of the Northeast and the Amazon Region that are viewed as aiding in the economic areas of these areas (fishing, reforestation, tourism);
- In the firm's own project of forestation and/or reforestation in the Northeast or Amazon Region provided that it complies with Article 18 of Decree Law 1.376/74;
- In tourism, fishing, forestation, and reforestation that are ap-

proved by the National Council for Tourism, Superintendency for the Development of Fishing, and the Brazilian Institute for Forestation Development.

Tax Exemptions. The tax on the Circulation of Merchandise (ICM) is usually not applied to the original sale of agricultural products, exportation of products to foreign countries, and the importation of merchandise under the draw back system. The Brazilian government has initiated a number of programs to stimulate the exportation of basic products and many nontraditional industrialized goods. These programs provide for exemption from

- Withholding income taxes (IR) in connection with the exportation of national products;
- Tax on industrialized products (IPI) in the case of exported products;
- Tax on circulation of merchandise (ICM) in the case of exported products;
- Tax on financial transactions (IOF) in connection with insurance operations associated with exports;
- Tax on transportation;
- Import tax (II) in the case of importation that is for the purpose of increasing export capacity.

Industrial Development Council (CDI). The Industrial Development Council (CDI) is an agency under the Ministry of Industry and Commerce that is responsible for analyzing and evaluating projects in connection with the granting of incentives. CDI is primarily concerned with improving the productivity and competitiveness of Brazilian enterprises, speeding up technological and administrative development, supporting development of engineering capabilities, developing new sectors oriented toward international markets, and expanding employment of the national labor force. The following is a listing of incentives that can be granted by CDI:

- Reduction of 50 to 80 percent of the import tax and the industrialized products tax levied on various equipment, tools, and instruments needed to complete priority industrial products;
- Exemption of the import tax and the industrialization tax when the project is approved by the President of the Republic;
- Credit to the purchaser of Brazilian equipment in an amount equivalent to the industrial product tax;
- Accelerated depreciation on Brazilian produced assets;

- Financial support from official financing agencies;
- Registration of foreign financing or investment;
- Granting of priority study examination by the Customs Policy Council (CPA).

The guidelines set by CDI for granting incentives are followed by the Customs Policy Council, the Bank of Brazil's foreign trade department (CACEX), and official credit sources such as BNDE. In recent years, CDI has been more strict in its granting of incentives in part because of Brazil's unfavorable balance of payments position. CDI is also very concerned that foreign firms use local sources of supplies for machinery, equipment, and other imputs. Under the "Law of Similars," CDI will only approve various incentives if Brazilian-made equipment does not exist.

Many firms have obtained good results in resolving source problems by working with the ABDIB (Brazilian Association for the Development of an Industrial Base). The ABDIB acts as a mediator and aids in resolving source problems involving the foreign investor and local Brazilian producers. The ABDIB has been quite effective in achieving agreement between foreign and local parties, and generally CACEX accepts the suggested arrangement in the judging of similars.

National Institute of Industrial Property (INPI). INPI is assigned the task of regulating the incoming flow of technology and assuring that such know-how is effectively utilized. Specifically, INPI is responsible for the registration of patents and trademarks owned by foreigners and for the approval of licenses that involve the utilization of foreign technical assistance by Brazilian operating firms. Through the activities of INPI, Brazilian officials feel that the present weak position of the Brazilian national company with respect to managerial and technical expertise will be corrected.[16] INPI has been strict in its approval of technology agreements because such agreements have often resulted in currency outflows from Brazil. In light of the nation's balance of payments position, it can be expected that INPI will continue in the future. In evaluation of a proposed agreement, INPI generally gives preference to

- Agreements involving companies having a majority equity position held by Brazilian interests;
- Agreements that hold the potential for aiding the expansion of Brazilian exports;
- Agreements involving know-how that is not locally available,
- Agreements in which the selling firm maintains a technical staff

capable of adequately implementing the new process or program in the Brazilian setting;

- Agreements that are associated with the development of identified high priority product groups;
- Agreements in which low levels of currency outflows or no outflows at all result;
- Agreements that have been identified as being beneficial to the Brazilian economy by other official government agencies.

Manpower, Management, and Research Training (Nonuniversity). The lack of capable management and technical personnel has been identified as a critical problem restraining economic development in Brazil. Various government agencies have introduced training programs to expand the capacity of the nation's labor force.

SENAI (National Service for Industrial Apprenticeship) is a private association of industries and industrial federations supported by the Ministry of Labor through a 1 percent levy on all industries.[17] In 1975, SENAI spent CR $927 million on 430,664 trainees throughout Brazil.[18]

FINEP has developed and conducts the PNTE (National Program for Executive Training), which is designed to increase the management capabilities of national firm administrative personnel.

CEBRAE (Brazilian Center of Management Assistance to Small and Medium Firms) is a nationwide system designed to improve and modernize small and medium firms in all sectors with respect to technology, organizational design, and management techniques. CEBRAE works through executives agencies called CEAGs (Centers for Managerial Assistance) that are located in twenty-two states. CEBRAE operates under the auspices of the BNDE, FINEP, and ABDE (Brazilian Association of Development Banks), and in cooperation with universities, research organizations, employers' organizations, and national, regional, and state planning and development agencies. Major activities of CEBRAE include regional and sectoral studies on various industries to determine specific needs and potential, training programs for top and middle management and for consultants, and support in preparing financial (loan) application to development banks. CEBRAE, with an operating budget of over $10 million and an organizational structure that facilitates diffusion of new techniques, has been an important source of technology for many Brazilian firms.

CENAFOR (The National Center for the Improvement of Trade Teachers) was established in 1969 by the Ministry of Education and Culture in collaboration with the United Nations. CENAFOR designs courses and provides technical assistance to other teacher preparatory institutions. A function similar to that undertaken by CENAFOR is now being carried out by the Institute of Human Resources of Getulio Vargas Foundation.

The Getulio Vargas Foundation (Rio de Janeiro) was founded in 1944, and trains accountants, economists, industrial psychologists, private and public administrators, political scientists, and human resource specialists. In addition, the foundation conducts economic research studies, publishes official statistics, and provides consulting services to public and private institutional groups.

Manpower, Management, and Research Training (University Level). Although the Brazilian government recognizes the critical importance of educational programs to the development of the nation's economy, and has expanded these programs in recent years, many observers have noted a growing gap between the demand for and supply of qualified white and blue collar workers. Not only are there insufficient numbers of persons being trained at the university level, but the quality of courses offered has been reduced as a result of the rapid expansion of educational programs.[19] In its current educational phase, the Brazilian government has committed itself to upgrading the quality of university courses, orienting courses more to the development needs of the community, and expanding the number of graduates, especially at post graduate levels.[20] One of the most severe problems faced by Brazilian universities is their inability to attract competent staff, in part because they are unable to offer salaries competitive with other sectors.

Political Factors Operative in the Brazilian Environment

General. The present military regime assumed power in 1964. While supporters of the government emphasize the progress made in the development and growth of the economy, critics point to a number of persistent problems in Brazil, including a decrease in income among the lowest income groups, a high level of unemployment, inadequate government attention to the development of the country's social infrastructure, political repression, denial of minority rights (black and indian), and increase in urban violence.[21]

Both viewpoints contain some truth. Brazil has made tremendous advances toward economic development but also has a number of serious social and political problems with which to contend. Such difficulties have not as yet, however, had a measurable impact on the operations of either domestic or foreign firms doing business in Brazil.

Brazil's Relations with the United States. The United States remains the most important trading partner for Brazil. During the last two years the United States has taken an increasing interest in the events taking place in Brazil.

Recent actions have seen the United States lifting tariff barriers on almost all Brazilian products. Such action has been predicted to result in a significant increase in the flow of Brazilian products to America.

Early in the Carter administration the controversy over human rights in Brazil resulted in strained relations between the two countries. Relations were further weakened by Brazil's agreement with West Germany to develop a breeder reactor.

The operations of U.S. public and private institutions have generally not suffered from disagreements between the United States and Brazilian governments.

Technological Development and the Brazilian Firm

Brazilian Firm Technology Users. According to a 1975 study, the major reason that Brazilian entrepreneurs seek sources of new technology is an interest in reducing costs and improving product quality to meet competition and to expand sales, particularly in international markets. Problems impending change in national firms relate to financial, technical, and personnel difficulties in obtaining and using technology.

National firms have moved to acquire technology from foreign sources to improve their competitiveness and expand sales. A recent survey of 127 firms operating in the Sao Paulo area showed that 60 percent had purchased foreign equipment during the last two years.[22] Eighty percent of the purchases were made by medium and large firms.[23] Of those firms surveyed that export goods, almost 90 percent had imported machinery and equipment during the two year period studied to assist in production. The competitiveness of Brazilian national firms, both domestically and internationally, is closely tied to modernization and adaptation of foreign technology.

From the above data it is clear that the small Brazilian enterprise imports less machinery and equipment, as well as other forms of know-how, and is less active in foreign markets than medium and larger firms. One explanation for this is that small businesses for the most part are closely tied to and run by family members who have little awareness of the existence and value of newly developed foreign technologies.[24] In general, most small businesses are locally owned. Companies having a high degree of foreign capital are generally large firms.

Some Major Factors Influencing Brazilian Technological Change. The development and utilization of new forms of technology require the use of high cost sources of funding. In the Brazilian market today, it is not uncommon to have 20 percent of the total cost of production attributable to costs of financing. Such high money costs make it difficult for Brazilian firms to invest in product or process technological improvements.

The lack of capable personnel—consultants, administrators, technicians, and skilled labor—is a major impediment to technical and economic development. Although the Brazilian government has launched ambitious training and educational programs, a high percentage of the nation's business executives maintain that a shortage (perhaps even a growing shortage) of competent administrative and technical personnel will exist in the future because government program efforts have been limited to date,[25] new technological innovations require the utilization of a progressively more highly trained staff,[26] and because the expanding Brazilian economy will demand more capable personnel than are being prepared by the Brazilian system.[27]

Except for the largest enterprises, the Brazilian firm has not been highly successful in hiring or keeping capable managerial and technical specialists. International companies have been able to attract a significant percentage of the top Brazilian labor force because they provide a more stable work environment, better opportunities for training and advancement, and more competitive salaries.

Cultural Patterns Having an Impact on Brazilian Firm Operations

Several empirical studies indicate that certain cultural practices do have a significant impact on the economic growth of many developing economies. Although generalizations are risky, a number of conclusions can be drawn in the case of Brazil that suggest the degree to which certain aspects have and will continue to have an impact on the growth of the national economy. These cultural factors relate to behavior in Brazilian businesses regarding the use of systematic approaches to making business decisions, the role of management development activities, the delegation of responsibility, and medium- and long-term planning and forecasting.

In contrast to American business practices, the Brazilian manager generally relies more heavily on intuition than on scientific management methods to arrive at important business decisions.[28] This decision process is primarily the result of the Brazilian manager's unfamiliarity with current management tools and the consequent tendency to adopt traditional ways of doing business. This lack of familiarity also tends to breed a certain amount of skepticism on the part of the Brazilian administrator toward the value of more modern methods of conducting business.

The Brazilian business firm and its management are closely intertwined with family relationships and politics.[29] The primary reason for this is the traditional dominance of the relatively small number of wealthy and powerful family groups in the various commercial sectors. Too often, management positions are filled by members of these elite families who have only marginal management experience and ability and who, lacking this background, do not understand the importance of manpower development to the firm's competitiveness, domestically and internationally.[30]

In most Brazilian national firms, the manager directs personally almost all aspects of company operations.[31] While there are a few large and progressive Brazilian national enterprises that utilize modern management methods, in the majority of cases the chief executive of the national enterprise only generally understands the intricacies of how to manage various aspects of production, finance, and marketing. Moreover, he tends not to delegate responsibility to those staff members who have extensive experience in these particular operational activities.

Most of Brazil's managers place greater emphasis on the attainment of immediate objectives than on long-range planning. This live-for-today attitude of many Brazilian businessmen reflects to a great extent their lack of understanding of the benefits of medium and long range planning activities. In summary, Brazilian public and private company officials as a group differ from developed country businessmen in their views toward the

- Utilization of systematic approaches to making business decisions;
- Importance of management development activities;
- Delegation of responsibility within the business entity; and
- Importance of forecasting and planning business operations in the medium and long term.

Foreign and Domestic Contributions to Brazilian Development Goals

Major goals of the Brazilian government are continued economic growth and development of indigenous productive capabilities. While local businesses do contribute to that economic growth, their capacities are not adequate for growth at a sufficiently rapid rate. Although the Brazilian government would like to "Brazilianize" the economy to the greatest extent possible, the more practical minded government officials have concluded that the nation is going to need significant, if not increased, foreign participation in the economy.[32] If Brazil is to realize its social and economic goals, it will need a considerable amount of foreign produced machinery and equipment,[33] foreign administrative and technical assistance,[34] and foreign capital.[35]

Two basic forces are at work in Brazil that will necessitate the country's use of foreign monies for a number of years to come. First, Brazilian national planners have adopted ambitious economic development goals, increasing the manager-scale development of infrastructure, and the development of heavy industry. This will require funds not currently available from local sources. Second, Brazil has had a severe currency outflow problem since 1974, a condition that further reduces the availability of local capital to support development projects.

The Brazilian national firm, nevertheless, has been charged by the government with carrying an increasing share of the responsibility for the country's development.[36] Many experts have pointed out, however, that the majority of Brazilian national companies are at a distinct disadvantage when it comes to competing with the foreign firm in both the domestic and international markets.[37] Specifically, most national firms do not use the most modern technology in production processes, have minimally effective administration and inadequately trained management and technical specialists,[38] and high labor turnover.[39] In addition, these firms typically conduct few, if any, training or development programs,[40] have limited research and development activities,[41] and operate under restrictive financial conditions.[42] A wide range of effective assistance programs will be needed to overcome these shortcomings and permit the Brazilian national firm to increase productivity levels and attain a truly competitive position in domestic and international markets. The Brazilian government has taken steps to strengthen the operations of national firms through programs aimed at

- Strengthening the economic viability of the national firm;
- Expending national research and development programs;
- Regulating the inflow of technology and foreign investment;
- Training and developing Brazilian personnel through university programs;
- Training and developing Brazilian personnel through federal, state, and local government agency programs;

- Assisting, through consulting and other services, the national firms' technical and administrative operations.

Although such efforts have improved the performance of the national enterprise, Brazil's limited human, capital, and material resources have reduced the effectiveness of the government programs. The following are indications of program weaknesses linked to national resource limitations:

1. *National Research and Development Efforts.*
   While almost all federal government agencies conduct some form of research and development activities, the Institute of Technological Research (IPT) and the Universities of Sao Paulo and Campinas are major sources of Brazilian-produced technology.[43] These institutions, and a limited number of others, provide Brazilian industry with some new technical know-how, but because their programs are limited many firms either operate without new technology or purchase know-how from abroad.[44]
2. *University Programs.*
   The number of Brazilians attending university programs has grown substantially during the last few years. Many observers of the Brazilian educational system note, however, that the increase in the quantity of the university programs has been accompanied by a decrease in quality.[45] One of the major reasons cited is lack of resource investment,[46] which, among other things, has kept salaries low and made it difficult to attract competent teaching staff. Inadequate teaching staffs at the university level translates into difficulties for Brazilian firms in locating sufficient numbers of competent administrative and technical personnel.[47]
3. *Federal, State, and Local Training and Development Programs.*
   As is the case with many Brazilian university programs, the lack of sufficient resources has limited the number of effective managerial training programs at the federal, state, and local levels. Because only 0.2 percent of the Brazilian firms have systematic training programs,[48] government-supported development courses must assume much of the responsibility for training the personnel of national firms.
4. *Brazilian Agencies Regulating Technology Flows and Foreign Investment.*
   Brazilian government agencies operating in these areas have met with only limited success. Numerous experts have observed that agencies such as INPI (National Institute of Industrial Property, the Brazilian Patent Office) do not have the technical capacity to carry out their assigned functions.[49] Weaknesses in agency opera-

tions have been attributed to a lack of human, financial, and physical resources.[50] The ineffectiveness of government agencies works against the operation of the national enterprises by (1) making it difficult for locally owned firms to acquire needed technology,[51] and (2) causing foreign owners of important technology to shy away from offering such know-how to Brazilian firms.[52]

5. *Financial Institutions Supporting Brazil's Economic Development.* To a certain extent, the large national enterprise has been able to satisfy its capital requirements through the use of Brazilian funding sources. A significant percentage of the country's firms, however, are experiencing high costs of capital[53] and problems of funds availability. Such difficulties have frequently caused firms to forego expansion and modernization.

In summary, the assistance programs of the Brazilian government focus on critical areas of business activity and generally make available to the entrepreneur resources that can aid him in improving the effectiveness of firm operations. However, as the preceding discussion indicates, many of the programs are not comprehensive enough to improve substantially the overall performance of the country's national business community. There is a pressing need in Brazil for creating programs able to capitalize on existing capabilities and resources for improving the development of Brazilian enterprises, thereby increasing their contribution to the country's economic growth.

PERU

General Policies of Government

Geoeconomic Base. Peru has a population of fifteen million living on 500,000 square miles divided between coastal deserts, high mountain ranges and underpopulated lowlands in the Amazon basin inland. Per capita income is about $500 per annum. The major cities of Lima (the capital, with a population of four million), Arequipa, and Trujillo are located in the coastal area and contain most of the industries. The mountain valleys are densely settled by Indian farmers and contain important mines. The Amazon region has lumber, tropical agriculture and oil.

Peruvian industries depend on imports for 45.5 percent of their inputs (1975 data). The principal imports in 1974 were

Table I-12. Peruvian Imports (1974)
(U.S. $ million)

Machinery	345	Fats	45
Iron and steel	137	Dairy products	41
Mineral fuels	123	Paper	35
Wheat	98	Precision instruments	35
Electrical machinery	91	Synthetic rubber	23
Resins and plastics	73	Miscellaneous chemicals	23
Organic chemicals	63	Inorganic chemicals	23
Motor vehicles	60	Books and newspapers	22
Tools and implements	20	Tanning and dyeing extracts	20
		Other	272

Source: U.S. Embassy, Lima, Peru.

To offset the annual cost of imports of about $1.5 billion, Peru exports copper, fishmeal, silver, zinc, sugar and cotton (these listed in order of importance).

Orientation of Government Policies. Peru's incipient democratic system came to an end in October 1968 when the Armed Forces took over the government. The program that has been implemented since that time by the Revolutionary Government of the Armed Forces has been characterized by the Peruvian Society of Industries as "statist, socializing, control-oriented and bureaucratic."[54] The government itself has described its posture as a third alternative to capitalism and communism and one that gives expression to the special identity of the nation. Its long-range program, "Plan Inca," grew out of prestudies carried out at the Institute of Advanced Military Studies in the context of antiguerrilla political warfare in the early 1960s. Its main features were a far-reaching agrarian reform as a way to secure rural tranquility by giving land and higher incomes to the small farmers, and a reform of industry as a way to secure labor peace by restructuring ownership and protecting job security.

The agrarian reform has expropriated holdings over fifty hectares of arable land equivalent, and given land to 300,000 families, albeit mostly in collective form. It has failed to raise their real incomes for a number of reasons. One of the most significant is the difficulty of rationalizing the agriculture of the overpopulated highlands without the benefit of a growing alternative job market in industries. The industrial reforms began in 1970 and have been formulated in a series of Decree Laws:

General Law of Industries D.L. 18350 (1970) and D.L. 19262 (1972) Industrial Community D.L. 18384 (1970) and D.L. 21789 (1977). The government has reserved for itself certain economic sectors, such as exploitation of natural resources, banking, and insurance. In the industrial sector, iron and steel, heavy machinery, basic chemicals, cement, and paper have been similarly restricted to state monopoly. The rest is open to private initiative, but has been greatly affected by other governments measures, including increasing competition from parallel state enterprises that enjoy credit and tax advantages.

The feature of the General Law of Industries that has had the greatest impact is the concept of the industrial community. All private industries with six or more employees and sales of more than 1 million *soles* ($12,500) are required to set aside 27 percent of their pretax profits to share with their employees. Of this, 10 percent was paid out in cash, 2 percent went to a fund for research and development (see ITINTEC below), and 15 percent was capitalized. The enterprise then transferred to the industrial community shares of equal value (15 percent of pretax profits). The industrial community was defined as the total work force (blue and white collar) of an enterprise. It held its shares in common and could not negotiate them. The shares were to be accumulated until 50 percent of ownership had been reached. This last feature in particular had negative impact on investment. The law was amended in early 1977; now only 15 percent of pretax net profit is paid out to the community, 11.5 percent is in cash and 13.5 percent is set aside in an account in the name of the community. The community decides on the investment of these funds in any of four different types of shares or bonds held individually by each member, except that 4.5 percent must be invested in shares in the enterprise, negotiable after six years. These may be accumulated up to the equivalent of 50 percent of equity, that is, 33 percent of total capital.

Foreign Investment D.L. 18900 (1971). Decision 24 of the Andean Foreign Investment Code, which establishes limits on foreign investment, has been applied enthusiastically by Peru. A fifteen-year limit was set in 1971 whereby all foreign-owned enterprises must reduce their ownership share to 49 percent (85 percent by 1974 and 65 percent by 1979). New investments are limited to 85 percent foreign equity, 70 percent five years after commencing production, 55 percent ten years later, and 49 percent fifteen years later. Profit remittances are restricted to 14 percent of the investment net of withholding tax.

Taxation. Corporate profit tax rates progress from 20 percent on the first 100,000 *soles* ($1,250) to 55 percent on profits over 1 billion *soles* ($12,500,000). Personal income tax rates start at 28 percent on 100,001 to 150,000 *soles*, and top out at 65 percent on income over 5 million *soles*. Businesses pay additional taxes—on equity (0.6 to 1.8 percent), real estate transactions, legal documents, sales and services, and registrations.

Tax incentives are offered under a number of headings (D.L. 18350; 19262; 21494). Tax relief is available from profit tax for reinvestment of net earnings and import duties. The amounts vary according to

- Type of industry: basic and certain manufacturing; manufacturing; production of nonessential goods; production of luxury goods; and
- Location (Lima/Callao; all other).

The range of exemptions is 65 to 99 percent of tax on reinvested profits and 8 to 100 percent of import duties. D.L. 20104 provides especially attractive incentives to the tourist industry based in part on locating outside of the Lima area.

Labor Laws. The Peruvian labor force is estimated at five million persons (31 percent of population). Unemployment has been growing and, taken together with underemployment, is estimated at 25-35 percent, but there is a shortage of skilled workers. Over 20 percent of the labor force is unionized. The average wage has been about 5,000 *soles* ($62) per month and is supplemented by legal social benefits amounting to another 60-70 percent (excluding profit sharing). Job security is protected by law; the only permissible grounds for termination are "grave cause" and closing of the enterprise.

Social Property Enterprise D.L. 20598 (1974). In an effort to provide socially countervailing forces to private enterprise in the sectors not preempted by other legislation, the government has formulated the concept of Social Property Enterprise. These are firms collectively managed by their workers and collectively owned by the sector as a whole. Profits are distributed partly on the enterprise level and partly on the level of the sector. Financing is provided by the National Social Property Fund, and supervision by the National Commission on Social Property (CONAPS).

Scientific and Technological Infrastructure

Peru has forty-four universities, of which twenty-two are government run and contain the majority of the 125,000 students (1973).

There are sixty-three affiliated research institutions sponsored by the government and ten private ones. Faculty members number 14,000, but only 115 are in the field of science and technology. In 1974 there were 77,150 graduates of Arts and Letters in the country, 24,400 in medical sciences, 11,649 in social sciences and law, and 9,655 in science and technology.

In 1973, 209 book titles were published. There are seventy-three libraries with more than 30,000 titles. Scientific research libraries traditionally have been attached to universities. The National Council of Peruvian Universities (CONUP) attempts to orient them to meet the requirements of national developmental policies. With the exception of the National Engineering University, most libraries are considered inadequate.

Management training on the university level is offered by ESAN (Advanced School for Business Administration) and the universities del Pacifico and de Lima. There are a number of intermediate management schools, such as IPAE (Peruvian Institute for Business Administration) or ADV (Sales Agents), which offer training in such disciplines as marketing, accounting, and personnel management. INAP (National Institute of Public Administration) was formed by merging ESAP (Advanced School of Public Administration), ONAP (National Office of Public Administration), and ONAL (National Office of Legal Assistance), and provides training for public administration.

In recent years, specialized entities of the government have developed their own information centers (ELECTROPERU for the electric power industry, PETROPERU for the petroleum and petrochemicals, MINEROPERU for mining, PESCAPERU for fishing). On the national level there are a number of institutions charged with the collection and dissemination of technological information:

CENIP—National Productivity Center Established in the 1950s as an adjunct of the Prime Minister's Office, CENIP sponsors seminars on productivity, time and motion studies, and acts as agent for the information and documentation services of OECD, UNIDO, and others.

ITINTEC—Institute for Research in Industrial Technology and Technical Norms Established in 1970 (D.L. 18350) as an autonomous agency of the Ministry of Industry and Tourism, ITINTEC receives 2 percent of the profits of industrial enterprises and is charged with providing them with technological services. The emphasis has been on establishing norms and standards and holding seminars. ITINTEC publishes information on technical processes, abstracts of articles from technical journals, a bibliographical series, a reference source entitled "Thesaurus," and a code of technical norms. In 1977 it opened an information center, which also offers films. Its 100 professionals are available for consultation on problems of special interest to firms.

CONITE—National Council on Foreign Investment and Technology The most important source of technology for Peru is the acquisition of foreign equipment and purchases of know-how under technology contracts. Control of foreign investments and approval of technology contracts are in the hands of CONITE (which took over this responsibility as of January 1977 from the Ministry of Industry and Tourism—along with a backlog of 3,000 cases including 800 technology contracts). Technology contracts are reviewed to conform with Article 19 of Decision 24 of the Andean Foreign Investment Code as regards specification of the project matter, breakdown of costs, duration (three to five years at present), and royalties. The government feels its review of these contracts has brought down the average royalty from 15 to 5 percent, in part by enabling the foreign exporters to qualify for foreign tax credits under the tax regimes of their domiciles.

Technical Training SENATI (National Service for Apprenticeship in Industrial Technology), which is financed by a 1.5 percent payroll tax, trains skilled workers for industries as well as foremen, quality controllers and trainers. ENTURPERU sponsors a hotel school. There are many private schools for secretaries, and commercial institutes.

Political Factors

Relations with the United States. Relations with the United States have been strained since the mid 1960s when the IPC issue emerged. The International Petroleum Company, a subsidiary of Standard Oil (N.J.), held exploitation rights to the Brea y Parinas oil fields near Talara. The Belaunde government, elected in 1963, was under constant political pressure from the opposition to abrogate the contract, which was considered very unfavorable to Peru, particularly its inclusion of certain rights of extraterritoriality, stemming from a colonial grant. The IPC, with U.S. government backing, stood on its legal rights. When in 1968, Belaunde finally reached agreement with it to buy out the fields, he was overthrown by the military, who insisted on outright nationalization. This led to the suspension of U.S. military and economic aid, and tension in U.S.-Peruvian relations. Peru turned for military aid to the Soviet Union, from which it has purchased significant quantities of tanks, and more recently, aircraft.

Socio-Political Change. The Statute of the Revolutionary Government, which was promulgated on October 3, 1968, gave the following as its objectives:

- Transform the structure of the State, making it more dynamic and efficient to facilitate governmental action;
- Promote higher living standards among the less favored sectors of the population, carrying out a transformation of the economic, social, and cultural structures of the country;
- Imbue the acts of government with a nationalistic and independent sentiment, focused on the policy of defense of the sovereignty and dignity of the nation;
- Increase morality in all fields of national endeavor, and reestablish completely the principle of authority, respect for the law and the rule of justice;
- Promote unity, concord, and integration of Peruvians, strengthening the national consciousness.

These policies were then implemented through a neverending series of decree laws which attempted to remake the nation in the image of the revolutionary statute. Yet these efforts of the military to reform the country and attain rapid social progress have been called "la revolucion de *El Peruano*"—revolution by the official gazette. The military have never succeeded in securing spontaneous popular support, in spite of all their efforts to do the right thing by the peasant and the worker. Some of the intellectual elite, the Social Progressists and the Christian Democrats in particular, shared the government's ideals and collaborated with it initially. Even they, however, found themselves to be little but distrusted advisors of the military. The bulk of the middle class has been in silent opposition.

The effect on business has been unsettling. During the eleven years since 1968, businessmen have never known what new law would appear in *El Peruano* in the morning. (Between October 3, 1968, and February 1, 1977, a total of 4,726 decree laws were published.) The general response has been to maintain positions, and suspend any plans for growth, investment, and hiring.

Economic Factors

Foreign Investment and Aid. The impact of the Peruvian General Law of Industries and the Peruvian interpretation of the Andean

Foreign Investment code has been negative on most foreign investment. Some of the companies that had major investments in the country (Grace and Co., Cerro Corporation) sought to reduce their exposure as quickly as possible. Yet, during the first half of the decade of the 1970s, Peru attracted more foreign investment than did most Latin American countries. This was due to two extraordinary items: the previously committed $300 million Southern Peru Corporation investment to develop the Cuajones copper deposits, and the search for oil in the Amazon region by a number of companies (Occidental Petroleum and others).

Following the military takeover, foreign aid was modest, due to the suspension not only of bilateral U.S. aid but also of a number of international bank loans that were blocked by a U.S. veto. In 1974, however, after the IPC dispute had been settled, international bank loans became readily available. By the end of that year, the Peruvian government public debt stood at $2,173 million. (In 1977 it was estimated to be well in excess of $5 billion, without knowing the extent of commitments for arms purchases.)

Balance of Payments. Unfortunately for Peru, just as foreign borrowing increased steeply, some of its most important commodity exports were faring poorly. Due to changes in ocean currents, the anchoveta fish were no longer abundant in Peruvian waters, and fishmeal exports declined disastrously. Where in the early 1970s, ten to twelve million metric tons were caught annually, the take in the mid 1970s was only two to three million metric tons. The prices of copper and cotton, and later sugar, were declining steadily. As a consequence, by 1977 the $700 million debt service was taking up 40 percent of the annual exports, and continued to increase. In 1975 alone the balance of trade was negative $1,113 million. Peru had to resort to a good deal of short term borrowing from private banks, at considerable cost. By 1977 foreign currency reserves were reported above zero only by minimal technical amounts.

Social Impact of Economic Trends. The Peruvian five-year plan for 1970-1974 had projected the creation of 294,000 new jobs in industry. In reality, less than 50,000 resulted. And the content of Peruvian inputs in Peruvian industrial products decreased from 62.5 to 54.5 percent. Nontraditional exports fell from 17 to 4 percent of the total. Consequently, the estimated rate of unemployment and underemployment was re-set at upwards of 35 percent. The Peruvian government could no longer afford to subsidize food prices as much as before. Prodded further by gaping government deficits, the inflation rate soared from the previous level of about 15 percent per annum to 46 percent in 1976. That year a 44 percent devaluation of the *sol* had to be carried out. (This was followed by a series of mini-devaluations until the exchange rate of *soles* to the dollar reached 80.88 by mid 1977, as compared with the rate of 43.38 maintained before 1976.)

Rising prices, falling real incomes, and lack of jobs have created considerable social pressures. There has been an increasing number of strikes, and rioting has erupted on occasion. In an effort to restore its credibility, the military government in 1975 replaced General Velasco with General Morales, a man associated with more moderate policies.

Physical Infrastructure

Transport. Peru has some 33,000 miles of roads. There are two main highways: the north-south 1,600 mile coastal Pan American Highway, and the east-west 523 mile Central Highway. The Marginal Highway running along the eastern slopes of the Andes is incomplete.

There are twenty-three ports on the Peruvian seaboard. The most important are Callao, which handles some three-quarters of all imports; the fishing ports of Chimbote, Paita, and Ilo; the iron ore shipping port of San Nicolas; the Amazon River ports of Iquitos and Pucallpa; and Desaguadero on Lake Titicaca. The ports are served by the Compania Peruana de Vapores, Prudential Lines, Lykes Bros. Steamship Co., and many others. There is a shortage of adequate storage facilities in the ports, and pilferage is high. There are 1,500 miles of railroads connecting population centers in the mountain zone with the coast. Airports in all major cities are served by a private Peruvian company, Aviacion Faucett. International passenger and cargo service is provided by Aeroperu, Braniff, and others.

Power. The installed electric power capacity of about two and a half million kilowatts relies about half on hydroelectric power and half on thermal sources. ELECTROPERU is expanding the generating facilities, which are primarily hydroelectric.

In 1974 Peru produced over twenty-eight million barrels of crude oil. Since that time, production from the Amazon jungle basin has been added to the coastal and off-shore established production areas. In 1977 the pipeline connecting these new fields with the coast went on stream, and Peru expects to be at least self-sufficient in oil by 1980.

Cultural Patterns of Business Activity

Family Patterns. Until the 1950s, the accumulation of wealth in Peru traditionally occurred among the oligarchic "40 families," who owned large extensions of irrigated land in the coastal valleys. Their agricultural operations were mainly export oriented, and focused on sugar and cotton. In the mountain zone other families owned feudal-style haciendas producing livestock. After World War II, foreign-controlled mining enclaves burgeoned, giving rise to new (supplier) industries. A Peruvian-owned fishmeal industry developed rapidly. Urban growth stimulated commerce, and the growth of import-substitution consumer industries. Through the banks, old wealth and new were shifting into industrial enterprises.

The behavior patterns developed through centuries by the land-owning families were carried into the new industries. Like the old extensive plantations, contemporary family-owned industries tend to be not oriented towards growth but towards security of investment and income. Certain job opportunities for family members is often a more crucial objective than is efficiency. There is a tendency to cartels and division of existing markets rather than aggressive development of sales.

View of the Private Sector. Unlike the generals and admirals of old, the new Peruvian military, certainly the colonels who were the spark of the revolt and have since been the mainstay of government operations, came from modest backgrounds, many from the provinces. They tend to view private business with suspicion and a considerable lack of understanding of how it really works. The legislation decreed by the Peruvian revolutionary government since 1968 has reflected this view. Large sectors of industry have been reserved for the state exclusively. There has been a consistent effort to devise alternatives to private business, the latest being the idea of the social property enterprise.

In April 1974 a Decree Law was passed providing for the social property firm and the social property sector. A social property firm can be founded at the initiative of one or more individuals, or by a government ministry or agency. Its directive committee is elected by its workers, as is its manager. In addition to wages, workers participate in profits. Start-up capital comes from the Government Development Finance Agency (COFIDE) and FONAPS (The National Social Property Fund). The firms could also issue "share-bonds." These instruments are shares in that their dividends are variable and bonds in that no ownership rights derive from them. The highest organ of the National Social Property System is the National Social Property Commission (CONAPS), whose chairman has ministerial rank. Twelve members represent government agencies, and three represent the social property sector assembly. It is not altogether surprising that by 1976 only one social property had been formed, and that in 1977, when the new "Tupac Amaru" Plan was promulgated, this idea was considerably downgraded. It is worth mentioning here only because it is the embodiment of the critique of the private enterprise by the military and their supporters. The social property enterprise was devised to assure worker participation in management and benefits. Income was to be distributed not on the basis of capital, but on the basis of work. Workers in one firm were to have equality with workers in any other firm in the sector.

Labor Relations. Like the management of industries, labor-management relations in Peru still reflect some of the attitudes and values acquired during the colonial and landowner-dominated eras. The *patron* is much more than the boss who organizes the work. He is the patriarch who claims privileges and demands complete obedience, and in return provides social security and counselling.

These attitudes are still very much in evidence in many of the small, family-owned businesses. (Decree Law 21435 governing small industries defines them as having sales under $500,000 and fewer than six employees. Of these, 3,500 were registered with the Society of Industries in 1977, 70 percent of the total.) The attitudes are held as much by the employees as by the employers.

The influence of labor unions has been felt much more in the larger enterprises. Here the classic adversary relationship has been the rule. The largest confederation of labor unions, the Confederation de Trabajadores Peruanos (CTP), has been dominated by APRA. The military government, for reasons of ancient vendetta as well as in defense of its current political interests, at first tended to favor the rival confederation, the Communist-dominated Confederation General de Trabajadores Peruanos (CGTP) which was smaller. More recently, as growing Communist influence, particularly among the miners, led to a series of unpleasant episodes, the government started to build a third confederation, the CTRP, initially based on fishermen's unions. The internecine struggles which developed have exacerbated labor-management relations. Union leaders felt they needed to strike ever more radical postures to compete with their rivals. There were 9,019 strikes between 1968 and 1975.

Some of the industries without a union still felt the impact of agitation backed by the government. SINAMOS (Sistema Nacional de Apoyo a Mobilizacion Social) has been the government organization charged with building grassroot support for the revolution by organizing peasant, worker, and community organizations of various types. SINAMOS agitation on the community level, reflecting the view that at best private enterprise is a temporarily necessary evil until the state and social property sectors could be built up, has affected some industries. More recently, however, SINAMOS has been toned down. Its leftist chief, the former Intelligence Director General Leonidas Rodriguez Figueroa, was exiled to Panama after Velasco's downfall.

Issues for Local Firms

Impact of Structural Reform Laws Under the military government, Peruvian businessmen have felt that they are operating in a climate of great instability. During phase I (under President Velasco, 1968-1975) a great many reformist Decree Laws were written by the COAP (President's Advisory Council) and approved by the Cabinet and the Revolutionary Junta (the chiefs of the three armed forces). This legislative overhaul was not carried out at once in the form of a general plan, but piecemeal over the seven years. As a result, every morning businessmen opened their newspapers with trepidation, fearful of the surprises which might affect them unfavorably.

During phase II (under President Morales, 1975 and on), while more moderate policies were being followed in general, there seemed less unity in the armed forces, and as a result the government line has tended to vacillate. For instance, during the general strike of 1977 the government authorized the firing of workers who broke collective bargaining agreements by the wildcat strike; later, however, businesses were required to take back many of these workers, who were then excused from their absence on a variety of technical grounds. The feeling of instability continued among businessmen.

A close second in importance among the problems perceived by businessmen has been the Labor Stability Law. In view of the extreme difficulty in laying off workers who become superfluous in a period of low sales, many industries have preferred not to employ them in the first place. The law has thus had the result of increasing unemployment while its objective was to increase job security.

With the exception of those cases where labor relations were traditionally poor, the Industrial Community Law apparently has not been a great problem to most established industries in Peru. Closely held businesses have been able to hold down the declared profits, and hence the share of the community, by a variety of devices. These include setting up separate service and sales corporations (commercial enterprises are not covered by the Industrial Community Law) and the concentrating profit in them by the use of judicious transfer pricing; or paying high compensation to family members. The law has, however, reduced investment in new industrial ventures; the 6,463 industrial enterprises with five or more employees in 1970 grew only to 7,783 by 1975. The fear of losing control of one's investment in the future has discouraged foreign investment even more.

Tax Incentives and Constraints. The General Law of Industries offered tempting tax benefits to investors. Profit reinvestment in particular has been stimulated by its provisions, which allow as much as one half of profits to be sheltered from the progressive business tax by being reinvested in the enterprise. This in turn has greatly strengthened the demand for foreign exchange with which to pay for the imports of capital goods financed by these reinvestments. In 1977 the Peruvian Industry magazine reported that only 0.05 percent of all Peruvians were paying the income tax, which indicates the availability of many ways to protect one's personal income from taxation.

The Law of Tourism, which offered additional tax benefits, has brought about lively investment activity in that sector. However, tax benefits offered to industries outside of the Lima-Callao area have not had the desired effect. In 1975 as in 1970, 70 percent of all Peruvian industries were still located in the capital city area. Apparently the many services and infrastructure available there have outweighed the tax considerations.

KENYA

General Policies of Government

Geoeconomic Base. The Government of the Republic of Kenya administers a land mass of 225,000 square miles, located on the eastern coast of Africa. The area is divided by the equater, and is characterized by highlands in the south and the southwest, where most of Kenya's farming is done, and desert plateaus in the north and northeastern two-thirds of the country. The population numbers approximately fourteen million, 85 percent of which lives in the southern and southwestern portions of the country, where the major urban areas of Nairobi, Thika, Nakuru, and the port city, Mombasa, are located. The northern part of Kenya is virtually waterless, suitable only for some livestock breeding, and is, not surprisingly, inhabited by only a few. The African population of Kenya comprises four major ethnic groups, each of which is represented by a major Kenyan tribe: Bantu (Kikuyu and Kamba tribes), Nilotic (Luo tribe), Nilo-Hamitic (Masai tribe), and Hamitic (Somali tribe). Non-African minorities in Kenya include Asians, primarily from India and Pakistan, Europeans, and Arabs.

Kenya has few known mineral resources, and its economy is based on agriculture, despite the fact that only 15 percent of its land is arable. A large segment of the population is engaged in subsistence farming, producing wheat, maize, and livestock for internal consumption. Primary cash crops and exports are coffee, tea, sisal, and pyrethrum. Growth in Kenya's manufacturing sector has been significant, maintaining a growth rate of approximately 10 percent over the past years, as opposed to growth rates in the agricultural sector averaging about 6 percent over the same period. Nevertheless, agriculture continues to be the mainstay of the economy, and accounts, along with forestry, for 29 percent of the GDP in 1976. The figure for manufacturing and mining for the same year is 14.8 percent.[55] The location of Kenya on the Indian Ocean makes it an important link to the countries in the African interior, particularly Uganda. Port facilities in the city of Mombasa and the trans-Kenya highway and railroad serve all of Kenya as well as African countries with no access to sea trading routes.

Kenya Before Uhuru. Kenya's history has been punctuated by the influence of various foreign groups. Indians and Pakistanis have long crossed the Indian Ocean for settling in Kenya, and Arab groups from the north have migrated into Kenya. Europeans also have a long history in Kenya. A vestige of early European presence is Fort Jesus in Mombasa, built by the Portuguese in the 1490s. In more recent times, other European groups, particularly Italians, have populated Kenya. It is, however, the British presence in Kenya that has had the greatest impact on recent Kenyan history.

Kenya came under British control at the turn of the last century, and in 1920 was organized as part of the Commonwealth with colonial status. In the 1950s, the Mau Mau uprising, whose roots were in the Kenyans' resentment of Europeans who continued to own much of Kenya's land, resulted in a delay in the granting of independence to the country. The movement leading to Uhuru, or independence, was led by Jomo Kenyatta, a member of the Kikuyu tribe who on December 12, 1963, became Kenya's first prime minister. In 1964, the Republic of Kenya was established, with Kenyatta as its first president, a post he held until his death in August 1978.

Kenya in its first one and one-half decades of independence has witnessed a political stability uncharacteristic of most postindependence African states. This has been due largely to Kenyatta's administration and careful balancing of tribal representation in his cabinet and throughout the government. Potentially disruptive disputes, particularly in the mid 1960s, were dealt with decisively by Kenyatta. A falling out with Kenyatta resulted in the forced resignation of Vice President Oginga Odinga in 1966. Other incidents that caused internal turmoil were the assassination in 1969 of Tom Mboya, who had been viewed as a successor to Kenyatta, and a plot in 1971 to overthrow the government. The strides Kenya has made in its economic growth have served to a degree to defuse disputes that may have arisen had economic problems been more severe. Nonetheless, the political stability of Kenya continues to depend on a careful balancing of interests of the dominant Kikuyu tribe, and the Luo and Kamba tribes that are seeking increased representation and control of the government.

Development Orientation of Kenyan Government. Kenya's development since Uhuru has been founded on the belief that both public and private sectors have an important role to play in economic and social development. In contrast to neighboring Tanzania, which has elected to follow a socialist pattern of state ownership of industry, Kenya chose a compromise course combining private resources and parastatal enterprises to fuel the productive processes of the country.

Immediately after Uhuru, the emphasis of the governmental development strategy was on education and industrial development. Kenya currently spends, and has spent for a number of years, 20-25 percent of the national budget on education. As a consequence of this amount being spent on education, industrial development was left to a large degree to the private sector, represented by both domestic enterprises and direct foreign investment. This industrial development policy emphasized import substitution initially, although in recent years, export promotion has become a primary goal of the government.

The third development plan (1974-1978) cited as priorities the improvement in income distribution and stimulation of jobs. Export promotion was viewed as a vehicle for achieving these goals, and became more important as a result of conflict among Tanzania, Uganda, and Kenya, the three members of the East African Community. As a result of growing tensions, the border between Tanzania and Kenya was closed in February 1977, which closed the gate on Kenya's major African export market. Since that time, the Kenya External Trade Authority was established in 1976 to promote Kenyan products abroad, and parastatal enterprises, such as the Kenya National Trading Corporation, have developed new export programs.

One of the major themes of the fourth development plan (1979-1983) is, according to Minister of Finance and Planning Mwai Kibaki, the "alleviation of poverty."[56] This focus is not a new concern of the government. The fact that the fourth development plan centers around this theme, however, is evidence of the government's attempt to formulate policies that will contribute to a greater equality in income distribution. Other issues that are emphasized in the 1979-1983 plan are the need to make industry more export-oriented, and to distribute greater amounts of government and international aid to marginal and semiarid areas of Kenya (the north and northeast regions).

Although there may be variation in priority areas in the development plans, the government has cited continuing development strategies that are to be fundamental in Kenya's development. These include

- Achievement and maintenance of a fast rate of economic growth;
- Increasing employment not only through faster economic growth but also through greater use of labour-intensive rather than capital-intensive means of production;
- Improvement of the living standards of the working poor most of whom earn their livelihood in rural areas on small holdings;
- Reduction of income differentials;
- Reduction in the divergence in wage rates between large and small firms for people in similar skills, to promote optimal growth;
- Making indirect taxation more progressive by imposing higher taxes on those items considered to be luxuries;
- Emphasizing economic production in those sectors which will increase exports and reduce imports.[57]

One of the government's primary mechanisms for economic and social development is its Kenyanization policy. "Kenyanization of Personnel in the Private Sector," issued by the government in 1967, holds that

> the primary responsibility of the Government of Kenya is to ensure that Kenya citizens will receive first priority in employment. This objective can be reached only when there are sufficient numbers of qualified personnel to make the vital decisions and to perform the many specialized and technical tasks which are required in a rapidly developing economy. This employment policy will not be allowed to impair the overriding objective of reaching the development targets set out in the development plan.[58]

Thus, while Kenyanization was not to be followed at the expense of economic efficiency and growth, there is continuing pressure in both private and public sectors to Kenyanize as much as possible. Part of this pressure is through encouraging private enterprises to develop training programs for employees to complement government-sponsored training centers and programs. This policy is monitored by the Kenyanization of Personnel Bureau, which requires that all firms wishing to employ foreign personnel submit an application to the bureau, and which requires that firms employing expatriate personnel file forms annually that summarize progress made in Kenyanization of their operations.

A second mechanism of the government for implementing development strategies is the parastatal enterprises. These are defined as enterprises in which the Kenyan government or local authorities designated by it own at least 50 percent of the equity capital. Established beginning in the mid 1960s, the parastatal enterprises serve to create employment opportunities and commercial opportunities for Kenyan citizens (through, for example, the Industrial and Commercial Development Corporation and the National Construction Corporation), to promote exports and reduce imports (Kenya Meat Commission), to push Kenyanization of the economy through training of Kenyans (Kenya National Trading Corporation), and to provide greater opportunities for Kenya's many rural small landholders (Agricultural Development Corporation and Agricultural Finance Corporation).

The Kenyan parastatal enterprises are linked to governmental ministries in their planning, and their activities reflect governmental development policies. At the same time, they are operating in the private sector, often competing with private companies for business, and, representing a far more serious problem, competing with private sector enterprises for scarce managerial and adequately trained technical personnel. For example, the Kenya National Trading Corporation's mission is to promote the Kenyanization of commodity trading in Kenya, that is, to train Africans in Kenya as local distributors of commodities in order to replace the Asian or other non-African traders. Thus, KNTC's purpose is to develop local private sector capabilities to operate effectively, which, theoretically, would make KNTC obsolete once that objective is realized. In short, the Kenyan development policies envision a cooperation between public and private enterprises in order to speed economic growth. However, because of manpower shortages in managerial positions, the public and private enterprises are, more often than not, competitors.

Orientation Towards Foreign Capital and Technology. Kenya has welcomed foreign investment since independence, and foreign capital and technology are important elements in the contribution of the private sector to the economy. Although the government favors and encourages joint ventures with 51 percent ownership by Kenyans, this policy is not rigidly enforced except in the mining sector. However, foreign investors are expected, in accordance with Kenyanization, to replace expatriates brought into Kenya or hired in Kenya with Kenyan citizens within a definite period, generally between one and five years.

The following summarizes regulations pertaining to foreign investment in Kenya:

Foreign Investment Protection Act of 1964 allows for the issue of a Certificate of Approved Enterprise by the Minister of Finance. This certificate authorizes the foreign investor to repatriate dividends, loan interest, and a portion of capital gained from sales. The certificate also protects the foreign investor as stated in the Kenyan constitution against compulsory acquisition of property by the government.

Foreign Exchange Control Act states that foreign exchange control is monitored by the Central Bank, that exchange rates and international transactions may be similarly controlled, and that strict import licensing is in operation, administered by the Central Bank and the Ministry of Commerce and Industry, in order to protect foreign exchange and local industry.

Immigration Act of 1967 outlines procedures for arrival and departure of foreign personnel for purposes of work or residence, and allows for monitoring of adherence to Kenyanization by foreign enterprises.

Foreign investment is particularly welcome in priority industries cited by the government, which include export-oriented industries where the value added is high, import substitution industries, agribusiness, labor-intensive industries, and hotel and tourism development.

Corporate tax is 52.5 percent of taxable profit for foreign or nonresident companies in Kenya, as opposed to the 45 percent rate levied on resident companies. A 20 percent withholding tax is levied on management fees, dividends are taxed at 15 percent, loan interest is taxed at 12.5 percent, and the capital gains tax is 35 percent. As incentives for foreign investors, tax deductions on capital expenditures (e.g., industrial buildings, truck and tractors, factory machines, and hotels) are allowed for newly established companies. Other incentives include a 20 percent investment allowance on the cost of industrial plants and machinery for companies locating outside the urban areas of Nairobi and Mombasa.

The procedure for investing in Kenya involves submitting a proposal to the Ministry of Finance and Planning, which, through its New Projects Committee, reviews the application and assesses the value of the project to Kenya's development. Upon approval by this committee, the proposal is submitted to the Capital Issues Committee, again in the Ministry of Finance and Planning, which arranges financial details of how the investment is to be made. Plans are then formulated for training of personnel and plant construction.

General Economic Outlook. The Kenyan model that draws heavily on the private sector, including foreign investment, has generally been regarded as an effective one for rapid economic growth. Kenya's progress since independence has been impressive, with a growth in GDP of approximately 7 percent every year (except for the 1973-1974 oil crisis year). GDP growth for 1978 is projected at 6-7 percent. Although this seems to indicate that the economy is moving in the same fashion as before, several factors in the Kenyan economy that are gaining importance warrant careful attention.

The problems caused by the oil crisis in 1973, particularly for the balance of trade, were alleviated in subsequent years by the rise in coffee prices. As Kenya's leading export, coffee accounted for 30 percent of total exports in 1976. In 1975, export earnings from coffee were approximately $170 million; in 1976, $233 million; in 1977, $690 million. The balance of trade deficit of $15 million in 1975 contrasts with the surplus of $35.5 million in 1977. However, the decline in the price of coffee and the drop in coffee crops projected for 1978 will cause a balance of trade deficit for 1978, estimated at $40 million.

The dissolution of the East Africa Community has had a negative impact on Kenya. The demise of the community was due in part to the feeling by Uganda and Tanzania that Kenya derived benefits from community membership far in excess of those enjoyed by Tanzania and Uganda. Tanzania was Kenya's major African market, and the loss of it has forced Kenya to seek new foreign markets. Until more effective export programs are implemented, however, the absence of Tanzania will be felt in the Kenyan economy.

A problem that is growing more serious in Kenya is inflation, running in 1978 at about 14 percent. Inflation since 1975 is as follows:

1975	*1976*	*1977*	*1978*
20 percent	17 percent	15 percent	14 percent

Although this pattern seems to evidence a healthy economy, some observers contend that trend is not so favorable. The 1975 rate reflects the quadrupling of the price of oil in 1973-1974, and represents in part an "imported" inflation. In contrast, the 1977 and 1978 rates reflect a rapid growth in the monetary supply resulting from the coffee boom, and represents an inflation that is more difficult to reduce.

Harambee Concept. *Harambee*, a concept introduced by Jomo Kenyatta immediately after Uhuru, is in a sense the development cry of Kenya. Harambee literally means self-help, and is a program based on the pooling of local resources for building local, particularly rural, institutions. Harambee has been extremely effective in supplementing government programs with private resources. For example, government funds for construction of rural schools are, as is always the case, limited, and have not been sufficient for the needs of the country. Under Harambee, citizens in numerous communities have pooled resources for construction of some 700 community schools. In many cases, the schools, once completed, have been taken over and supported by the government and administered by the Ministry of Education. Through Harambee, other projects such as construction of churches and water projects have been executed in rural areas, and the Harambee movement can be said to have contributed significantly to the spreading of Kenya's development beyond the major urban areas. A persistent problem with Harambee, however, is the increased politicization of the movement, with local politicians seeking support from Harambee funds or with use of Harambee funds for narrow or individual interests.

Future Trends and Problems. A major, if not the major, problem facing Kenya today is the shortage of skilled manpower, both technical and managerial. The paucity of qualified personnel to fill managerial and technical positions means that private companies compete with each other and with parastatal enterprises for scarce manpower. This causes significant problems for the Kenyan development model that is based on private/public cooperation. Managers, particularly in the parastatal enterprises, fear loss of personnel to higher-paying jobs in the private sector or poaching of personnel by other state-owned enterprises. At least one general manager of a parastatal enterprise indicated that this likelihood of loss of personnel has discouraged him from instituting management training programs, which, unfortunately, is precisely what the organization needs.

Other major problems facing Kenya in the future stem largely from the rapid (3.5 percent) population growth. Fifty-four percent of Kenya's population is under the age of fifteen, and the problem is exacerbated by the absence of effective government programs to deal with the increasing population.

This population growth rate has numerous implications for Kenya and its economic development. Agricultural production must be increased, and programs should encourage the reclamation of marginal lands and better use of land currently under cultivation. Industrial production must be increased, which requires foreign technology and capital in addition to local input. However, to continue to attract and hold foreign investment, infrastructures must be strengthened, particularly in the areas outside of Nairobi and Mombasa that the government is attempting to develop. Water is scarce in much of Kenya, and water development is a prime concern of the government at present. Jobs must be created—estimates are that by the year 2000, Kenya will need fifteen million additional jobs. Creating jobs and developing manpower to fill those jobs requires innovative training and educational programs for all levels of work (technical and managerial). The gap between rich and poor in Kenya will undoubtedly grow wider before policies to distribute wealth more equitably are or can be implemented, and may be a source of growing frustration and tension in the country. Kenya's dependence on foreign sources for oil means that balance of payments problems could increase unless export programs are effective. Kenya's export promotion policy emphasizes products manufactured in Kenya in order not to be dependent only on commodity exports such as coffee and tea, whose prices fluctuate. However, markets for these goods must be located and developed rapidly in order to reduce potential balance of payments deficits.

Reports of corruption among government personnel persist, and may or may not be a problem for Kenya in the future. Kenya's political stability has been assured in the past through the rapid economic progress the country has made and through the balancing of tribal interests. There is in Kenya a rapidly growing class that feels it has much at stake in the present system, and that will probably not risk its current position solely in order to eliminate corruption. However, the growing population and the increased perception of a widening wealth distribution gap may result in increased disaffection with current government practices and may necessitate more rapid programs for rural development and more equitable income distribution.

Scientific and Technological Infrastructure

Formal Educational Institutions. As noted earlier, a basic tenet of Kenya's governmental policy has been the importance of an educated population, and this importance is reflected in the 20 percent of the national budget that is allotted to education. In 1963, total enrollment in primary institutions was 891,553. Enrollment in primary schools in 1976 was 2,894,617. Of these, 2,388,270 were ages six to twelve, and this number represents an estimated 86.7 percent of the Kenyan population in that age group.[59] Enrollment in secondary schools for the same year was 280,400, reflecting an increase of 24 percent over 1975, which was possible because of the additional 120 secondary schools that opened in 1976. Enrollment at the University of Nairobi, founded in 1970, was 5,250, and at Kenyatta University College, a school founded in 1972 to train teachers, enrollment was 1,127.[60]

With the emphasis on formal education, there has been too little attention paid to technical and vocational training in Kenya. The result is that there are too many people with academic training and too few available for skilled technical/vocational jobs. Several observers indicated that a significant problem is the rising expectations of university graduates regarding the possibilities for top level positions. While Kenyans with university degrees were in great demand immediately after independence was won in 1963, there are at present too few administrative and managerial positions open for the number of inexperienced graduates expecting to find that kind of work. It is expected that of the 1800 graduating in 1978 from the University of Nairobi, 13 percent will remain without gainful employment for at least three years.[61]

Technical and Vocational Institutions. There are two major polytechnical schools in Kenya, located in Nairobi and Mombasa, with combined enrollments of 4,240. In addition, Utalii College offers training for the tourism industry. Courses are offered as well by a school run by the East Africa Power and Lighting Company, by the railway and post office training schools, and by the Kenya Institute of Mass Communication.

Under the Harambee movement, a number of primary and secondary schools were constructed and continue to provide formal education for some 30,000 students. Established with local community funds, many of these schools are being taken over by the government and assisted by the Ministry of Education. In the last several years, 200 small technical schools called village polytechnics have been set up under the Harambee movement. These provide training in craft skills such as bicycle maintenance and repairs and basketmaking. In addition, institutes in Kiambu, Murang'a, and Kirinyaga have been established to provide training in somewhat more sophisticated technical skills than those found in the village polytechnics.

Foreign Sources of Technology. There are in Kenya a number of foreign influences through which Kenya is exposed to foreign development techniques. Nairobi is a key city in East Africa and is rapidly becoming a major conference center. The East African Community was headquartered in Nairobi until its dissolution in 1977. The United National Environmental Programme is headquartered in Nairobi, the implication being that many countries are represented in Kenya that would not maintain consulates for only bilateral dealing with Kenya.

Foreign aid comes to Kenya both through bilateral agreements with national governments and through international organizations such as the World Bank and the International Monetary Fund. In 1977, foreign aid amounted to $112 million, and is expected to be between $175 million and $200 million per year throughout the fourth development plan.

Kenya has depended on foreign direct investment for much of its technology for industrial development. Over 100 U.S. companies have offices in Kenya, and even more European companies operate in the country. In addition to providing hardware, these companies generally provide training for Kenyan employees as a result of governmental encouragement to do so. Indeed, after a foreign investment proposal is approved by the New Projects Committee and the financing arranged by the Capital Issues Committee, some training of manpower is required even before plant construction.

A number of voluntary and research organizations are also present in Kenya and represent another source of technology. The International Executive Service Corps and the British Executive Service Overseas are but two of these organizations providing assistance to local businesses. The Canadian International Development Research Center and the Ford Foundation are examples of foreign nonprofit organizations that can be tapped as sources of technology.

Government Programs for Technology Assistance. In addition to the formal and technical educational institutions noted above, the Kenyan government has developed programs to provide training and managerial and technical assistance to local businesses.

The Management and Advisory Training Centre was established by the Ministry of Labor in the mid 1960s with the assistance of the United Nations Development Programme. The purpose of the center is to promote more effective management in Kenya and to assist new and established businesses to resolve management problems. The center offers courses to businessmen and provides some consulting services as well. Also operating under the auspices of the Ministry of Labor is the Industrial Vocational Training Centre, which provides vocational and technical training through regular courses.

A program that has special significance for small businesses is the Kenya Industrial Estates. KIE is a subsidiary of the government-owned Industrial and Commercial Development Corporation, and assists in the development of small industry. In many KIE projects, an investor provides capital, staff, and management for construction of a small plant. KIE provides assistance in locating technology, raw materials, and consultants, and may also assist in financing of the project. Production of bolts or hinges, stainless steel knives and forks, and office supplies are examples of KIE-assisted projects.

The Industrial Survey and Promotion Centre is charged with outlining possible investment projects that could be undertaken by foreign investors. Because of the research it conducts on current production capacities and projected needs in specific industries, it represents a source of information that can be tapped by local business. However, most of its research relates to large industrial developments, and is not necessarily the basic general management information and assistance many local businesses require. After the demise of the East African Community, Kenya inherited the Industrial Research Organization, which was set up under the community to study industrial development needs of the three member countries.

Finally, the parastatal enterprises play a significant role in the technological infrastructure, particularly with regard to industrial development. The Industrial and Commercial Development Corporation, the Industrial Development Bank, the Agricultural Finance Corporation, the Agricultural Development Corporation, and the Kenya National Trading Corporation are several examples of parastatal enterprises that are operating throughout the country and that are consequently involved to varying degrees in assessing Kenya's current capacities and future needs for industrial development. As noted earlier, these enterprises are set up as profit-making institutions and operate in the private sector in the sense that they compete with private businesses in areas where they do not hold a monopoly position. However, because of the extremely high turnover rate, they serve in addition as mechanisms for diffusion of technology. The shortage of qualified managerial personnel forces public and private sector enterprises to hire managers out of other organizations. These personnel, of course, carry managerial techniques with them to their new positions.

The National Council on Science and Technology was established in June 1977 with the purpose of reviewing Kenya's technology needs and formulating a national technology policy and recommending priority areas for technology acquisition. The council was established with the assistance of UNESCO, and one of its first duties has been the preparation of the Kenya national paper on technology for submission to the United Nations Conference for Science and Technology for Development. All permanent secretaries of the ministries of the government of Kenya sit on the council. The council works in conjunction with existing government offices, such as the research wing of the Ministry of Agriculture, the Kenya Industrial Research Organization, the Kenya Institute of Administration, and the Ministry of Commerce, in order to set up programs to channel foreign technology into the government-cited priority areas of promotion of industrial development and creation of employment. Because the council is still quite young, it is difficult to assess its effectiveness in formulating a national technology policy and overseeing the implementation of that policy. However, some observers in Kenya fear that, without a relatively rapid development of policy, the council will become merely a clearinghouse for proposals for foreign investment instead of the more research-oriented organization that is required in order to assess Kenya's pressing technology needs for rural and industrial development.

Other Sources for Technical or Managerial Assistance. There are several other institutions in Kenya that represent additional sources of assistance to local firms. The Kenya Institute of Administration and the Kenya Institute of Management provide training in general managerial and administrative techniques. The Christian Council of Kenya provides assistance to both management and labor groups.

There are in Kenya only a few industry or professional associations that provide much managerial or technical support to local firms. In many instances, these associations are used as lobbying groups to exert pressure on politicians and to encourage implementation of policies that will favor the particular industry. In far fewer cases are these organizations used as training centers or as forums where local businessmen can discuss specific problems and exchange ideas with other business owners. General business organizations include the Kenya Federation of Employers, the Kenya Association of Manufacturers, and the national Chamber of Commerce. The Association of Economists was established in 1977 as a forum for industry, academic, and government economists.

As the African-owned segment of the private sector grows and displaces the expatriate firms, which in many industries still dominate the market, it is likely that new industry associations will be formed. Surveyors of Kenya was established only after several African-owned firms began to operate in the real estate development industry. The Kenya Institute of Certified Public Accountants was established recently, only after there was a demand from African-owned accounting firms for a national institute to certify Kenyan accountants. Before the founding of KICPA, accountants in Kenya were certified under British or U.S. systems. In the early stages of development of African-owned industry, it is not surprising to find very few professional associations that provide support for local firms. Indeed, these associations in their early years seem to be held together only by the commitment of one or two dominant African-owned firms.

A final source of foreign managerial and technical aid is the missions of national governments. U.S. AID, for example, cooperates with major businesses in specific projects and can be tapped as a source of information.

Political Climate for Local Business

Internal Political Environment. As noted earlier, the political environment in Kenya has been remarkably stable since Uhuru. While some African countries have fervently adopted political ideologies, Kenya's political climate primarily reflects the national economic goals of industrial development and employment creation. The policies followed by President Kenyatta are likely to be continued under the administration of Daniel arap Moi, who, after Kenyatta's death, was elected president in early October 1978, having served as vice president under Kenyatta. President Moi is a member of the small Kalenjin tribe, and it is expected that this will serve to defuse political instability that could stem from his belonging to one of the two major tribes that historically have vied for dominance in the country. President Moi now heads the Kenyan African National Union (KANU), which was founded by Kenyatta and which, since the Kenya People's Union was banned in 1969, has been the only political party.

The development goals of the country and the internal political climate have provided a healthy environment for local business. The government has encouraged local business development through creation of the parastatal enterprises and through Kenyanization. However, African-owned business development has not proceeded as rapidly as might be expected, given the private sector's role in Kenya's development, due to shortages of qualified manpower.

Relations with Neighboring Countries. Kenya shares borders with Tanzania, Uganda, Sudan, Ethiopia, and Somalia. The good relations with Tanzania and Uganda for a number of years following independence of the three countries is evidenced by the formation of the East African Community in the 1960s. Over the past several years, however, the relations have been strained among the three states. Idi Amin's rule in Uganda resulted in the breaking of ties between Tanzania and Uganda. The strained ties between Tanzania and Kenya resulted in the closing of the border in late 1977, and ultimately to the dissolution of the community.

The more recent development in the Horn of Africa causes particular concern for Kenyans. Kenya generally would support Somalia in its claim to the Ogaden region of Ethiopia because Somalia's pro-Western stance is more in line with Kenya's than is the more pro-Communist bloc stance of Ethiopia. However, the Somalis reportedly believe they have claim to a region of northern Kenya, and victory by the Somalis in the Ogaden question might, the Kenyans fear, encourage the Somalis to act on that claim. Kenya is aware of problems that could result from its geographic position and relations with neighboring countries (decrease in number of tourists, loss of Tanzania market), and is attempting to compensate for these problems through aggressive tourism promotion campaigns and export market searches.

Relations with the United States. Kenya's pro-Western position and development policies favoring both private and public sectors have contributed to generally good relations between the two countries. The U.S. government is represented by its embassy and through a number of U.S. AID projects. As noted earlier, many U.S. companies—manufacturing, insurance, food processing, banking—have operations in Kenya, both in the major industrial areas of Nairobi as well as in rural areas that the government is trying to develop. Ties between the two countries are strengthened through the presence of U.S. private foundations and volunteer groups.

One area where the tie is not particularly significant is in military assistance. Kenya supports only a small civil military, spending no more than 6 percent of the national budget on defense. What military assistance Kenya does receive (small air craft, advisory and technical assistance, uniforms) comes primarily from Great Britain and the United States.

Physical Infrastructure

Mineral and Energy Resources. Kenya's mineral reserves are minimal and mining is not a major industry in Kenya. Kenya has no oil reserves and is, therefore, dependent on imports for fueling its development, making its economy vulnerable to increases in oil prices. Table I-13 indicates that over $171 million of Kenya's imports came from Iran in 1976, and there is little question as to the composition of those imports. Kenya is also quite poor in water resources, making development of hydroelectric plants impossible. The more fundamental problem at present is simply providing water for the large portion of the country that has little or no water sources. Lack of water has proven a significant problem for both foreign and local business, particularly in areas outside of Nairobi. Construction of new game lodges for developing the tourist industry has, for example, been thwarted because of lack of water or roads accessing sites.

Table I-13. Kenyan Trade (1976)
($ million)

	Imports	*Exports*
United Kingdom	192.5	90.2
U.S.	59.2	46.0
Iran	171.0	–
West Germany	102.5	106.2
Tanzania	–	83.5

Roads and Railroads. Kenya has 27,000 miles of classified public roads, not all of which are paved. A major highway extends from Mombasa into Uganda, and another connects Nairobi and Arusha in Tanzania. The Trans-Africa Highway now under construction will begin in Mombasa, and cross through African states between Kenya and Nigeria.

Rail travel in Kenya has been possible since late in the nineteenth century. Nairobi was founded during construction of the railroad less than a century ago. The 1,700 miles of rails connect Mombasa to Nairobi and Kampala, with branches into Tanzania and Uganda.

Port and Air Facilities. Mombasa, with its Kilindini Harbor, is the largest port on the coast of East Africa, and has historically been an important trading center. Mombasa is the entry point for goods into Kenya and Uganda. The development of several industries, particularly the tourist industry, has required improving air facilities as well. Nairobi's airport is now serviced by twenty-five international airlines, and its new passenger and freight terminal is the largest of its kind in Africa. The Mombasa airport was improved in 1977 to meet international standards.

Dominant Industries and Markets

Dominant Industries. Agriculture continues to be the dominant industry in Kenya, and accounts for over 25 percent of GDP. The government continues to emphasize agriculture as a primary industry, and is looking for foreign investment particularly in the areas of meat processing, dairy products, and canning and food processing. Growth in the manufacturing sector has, as noted earlier, been more rapid than in the agricultural sector, and Kenya is looking as well to assistance in building its manufacturing businesses. Tourism has always been a relatively active industry in Kenya simply because of the country's splendid variety and supply of animals. In recent years, however, the government of Kenya has become more concerned with developing other aspects of Kenya's tourist trade. New facilities for accommodating tourists include new lodges in the game parks, new hotels in Nairobi, the Kenyatta Conference Center in Nairobi, and improved airport facilities in Nairobi and Mombasa.

Dominant Markets. Traditional exports from Kenya have been commodities such as coffee, tea, sisal, pyrethrum, and cashews. Kenya's largest export market outside Africa was Great Britain until very recent years, when increased demand for coffee from West Germany pushed that country to the top of the list.

Within Africa, Tanzania was Kenya's largest market until the closing of the border. The following illustrates trade patterns with selected countries for 1976.

Another export from Kenya has been game trophies and animal skins, although much of this activity has been illegal. Poaching, or the illegal killing of animals for sale of skins, became such a serious threat to some species that former President Kenyatta announced in late 1977 that the sale of trophies in Kenya would be banned after March 1978. Kenya is, however, attempting to develop through the Kenya National Trading Corporation international markets for tanned animal hides.

Issues Affecting Local Business

Tax Incentives and Constraints. The corporate tax for national firms is 45 percent. Incentives for specific industries and for location in areas outside of Nairobi and Mombasa, as discussed with regard to foreign investment, apply as well to Kenyan firms.

Until the early 1970s, there were virtually no restrictions on importation of technology. In order to protect local industry, however, a 30 percent tax on imported technology was imposed in 1971 and has been in effect since that time.

Managerial Practices. Although at present companies from many countries operate in Kenya, the managerial practices of Kenyan businessmen reflect primarily systems inherited from the British. Organizational structures follow the patterns of British organizations, with a corporate secretary as one of the key figures in the company. One observer noted that the relationship between managers and employees is akin to the human relations approach that is more frequently found in British than in other northern European companies.

Despite these traditional ties to British management systems, the influx of the foreign managers and the growing numbers of Kenyans who have worked or studied in countries other than Great Britain are contributing to changing patterns. For example, one Kenyan real estate developer has adopted certain U.S. real estate practices (use of standardized leases instead of preparation of a separate contract for each leasing agreement), and feels that use of techniques other than those of the British will enable the Kenyan business community to be more independent of the established British firms and therefore develop more rapidly.

Labor Activity. COTU, the Central Organization of Trade Unions, is the organization representing the fifty-two labor and trade unions in Kenya. The labor unions in general are quite weak because of the lack of financial resources and because of government policy that has not encouraged the development of strong unions. COTU's activities consist primarily of publishing statements in the press. Being a vocal organization is perhaps its compensation for having little real power.

Hiring patterns of the Kenya business community reflect a frequent use of casual labor. "Casual" is the term applied to employees hired for short periods, generally one to three months, for low-skilled labor such as cleaning and messenger or delivery services. The casual labor force is generally unorganized, and laws relating to casuals are less stringent than those regulating the permanent employees. Parastatal enterprises as well as private firms use casuals—in one parastatal, casuals number 300 out of a total employment of 600 in the firm.

Labor laws in Kenya allow firing of personnel only for reasons of redundancy of personnel, for disciplinary action, or for inefficiency in performance. Disputes over firing or other labor issues can be handled through discussion between the union and management or through the Ministry of Labor's Industrial Court. The majority of cases are heard before the Industrial Court, due in part to the weakness of the unions in negotiating directly with management.

Manpower Sources. A significant problem for local businesses in Kenya is attracting and holding personnel that are qualified and trained in business skills. Migration patterns in Kenya are such that there is an abundance of labor in the urban and business centers, but that labor is unskilled and, unless trained, unable to contribute to local business development. A second category of manpower consists of recent university graduates, trained, the theory posits, for white collar jobs. Indeed, the shortage of university graduates of a decade ago meant that all Kenyan graduates were placed in relatively significant positions in private and parastatal enterprises, the result of which is an attitude among many in Kenya that a managerial position is the least that can be expected. However, few of these graduates have the practical skills or experience required for the management needs of most Kenyan businesses. Many local businessmen have hired expatriates living in Kenya, or have traveled to other African countries and Great Britain to hire, for example, qualified accountants. Businesses eventually will use this practice less frequently under pressure to Kenyanize and as the supply of adequately trained Kenyans increases. In light of Kenya's pressing need for experienced managers and administrators, however, "eventually" remains in the not too near future.

TANZANIA

General Policies of Government

Geoeconomic Base. The United Republic of Tanzania, comprised of Tanganyika and the islands of Zanzibar and Pemba, has 365,000 square miles and is almost as large as Texas and New Mexico combined. Of the total mainland area (363,950 square miles), more than 23,000 square miles are inland water. About 60 percent of the country is not arable because of tsetse fly infestation and lack of water. Most of the arable land is located in the northern and southern highlands, along the coast, and in the western lake area.

According to 1976 statistics, Tanzania's population is 15.4 million and growing at an annual rate of 2.7 percent. The population density is forty-two people per square mile, but the distribution is uneven, ranging from three people per square mile in arid areas to 133 in well-watered highlands to almost 350 on Zanzibar. More than 90 percent of the population lives in rural areas. The largest cities include Dar es Salaam, the capital (500,000); Zanzibar Town (80,000); Tanga (60,000); Dodoma (35,000); and Mwanza (35,000).

Tanzania is primarily an agricultural economy, and governmental policies are designed to maintain this emphasis. According to latest figures, there are approximately 460,000 wage earners in the entire country, and these are employed in textiles, agricultural processing, light manufacturing, mining, refined oil, and cement. In addition, a growing number of workers are employed in administration and services. Because statistics are not current or accurate, it is difficult to obtain more than an approximation of absolute numbers or a trend in development. Another complicating factor is the current policy of government takeovers and transformation of private firms into parastatals in key industrial sectors.

What is clear is that the private sector, despite the rapid expansion of parastatals, still accounts for about one-half of production. Tanzanian industry accounts for roughly 10 percent of GDP, which was measured at $2.4 billion in 1976. Industrial wages are estimated at about $800 annually, compared to GDP per capita of $160. This makes Tanzania one of the world's poorest countries according to U.N. definitions.

A 1973 survey showed a little more than 500 manufacturing establishments with ten or more employees. By 1976, there were 330 parastatals, of which 175 were engaged in business. With the exception of an untapped hydroelectric potential, unexploited iron and coal, and dwindling gem stone and gold mines, Tanzania is not endowed with many natural resources. Its monetary economy is based on a number of primary and semiprocessed agricultural exports such as coffee, cotton, tea, sisal, cloves, cashew nuts, and pyrethrum. Industrialization is aimed primarily at import substitution and processing agricultural commodities.

History. Despite the anthropological studies done by Dr. Leakey in the Olduvai area, an accurate history of African people in what is now Tanzania is more or less unknown. This lack of information is especially true about the mainland interior. It is known, however, that a number of tribes—Bantus, Nilotes, and others—had well-organized societies and controlled much of the area by the nineteenth century when Arab slavers and European explorers and missionaries penetrated the area. On the coastal areas, foreign influences were felt from the eighth century, mainly of Arabs, Persians, Indians, Portuguese and Chinese, whose primary interests were as traders.

German colonial interests were initiated in the mid 1880s with the establishment of the German East Africa Company. This occurred at the time that the British and German governments negotiated spheres of influence in the interior of East Africa. In 1891, the German government took over administration of the East Africa Company. The Germans were responsible for establishing the beginning of an economic infrastructure—plantation economy and development of certain cash crops, railroads, roads, harbors, buildings, and the like. This arrangement ended with the Treaty of Versailles when Tanganyika became a trusteeship of the British. After World War II, Tanganyika became a U.N. trust territory also administered by the United Kingdom. From that time, Tanganyika moved gradually toward self-government and independence in 1961 under Julius Nyerere's leadership.

Zanzibar was dominated from time to time by Arabs, Portuguese, Indians, and Persians. Valued as a trading center, Omani Arabs, utilizing forced labor of the island's African population, developed clove plantations that still produce Zanzibar's main export. The island was also a center of slave trade. Zanzibar eventually came under British influence, gradually moved toward independence, and became a constitutional monarchy under the Sultan in late 1963. In early 1964, the African majority revolted against the Sultan and a new government was established. In April 1964, Tanganyika and Zanzibar were politically united with the latter retaining much local autonomy.

At first, Tanzanians simply moved into key positions in institutions inherited from the colonial legacy. In time, President Nyerere evolved a conception of a society that would bypass some of the problems of modernizing societies. His agrarian socialist model aims at both growth and equity with limitations on the accumulation of material wealth but capable of providing an adequate level of material existence. The Arusha Declaration of 1967 and the Mwongozo guidelines of 1971 laid out the details of this new socialist experiment in some detail and put heavy emphasis on self-reliance and participation through workers' councils, a degree of self-management, and collectivation (Ujamaa villages).

At the time of independence, Tanzania had a very underdeveloped industrial base. Only about 20,000 people were employed in manufacturing, and industry contributed less than 5 percent of GDP. Foreign consulting firms and international organizations were relied upon for devising domestic production strategies. Import substitution and processing of local materials for export were seen as short-term and long-term strategies respectively. With more than 90 percent of the population already rural and eventually living under the Ujamaa concept, industrialization projects are secondary to agricultural development. There is nonetheless no lack of foreign assistance presence in Tanzania. Roughly $20 per capita in current foreign grant assistance testifies to the high interest of donor countries in Tanzania's socio-economic experiment.

Foreign Technology and Capital. Tanzania's critical shortage of trained manpower, infrastructural deficiencies, emphasis on village agriculture, and policies to promote small and medium scale import substitution industries make it a relatively unattractive market for foreign direct investment. The government does not explicitly prohibit foreign equity in industrial projects but insists on majority ownership in joint ventures. Policies on the importation of foreign capital for investment are not clearly stated. Terms and guarantees are always negotiated for each project, and they vary by sector and priority of development area. Laying the groundwork for this flexible but unclear policy are the Foreign Investment Protection Act of 1963 and the Short Guide for Investors, 1967. Neither Tanzania's socialist ideology nor restrictive legislation frustrate foreign direct investment as much as do market factors mentioned above. Other events minimizing the potential role of foreign investment in industrial development and negatively affecting market dynamics and expansion are the border dispute (and closing) with Kenya and the dissolution of the East African Community.

In terms of external economic relationships, Tanzania apparently feels that it has greater control over the direction of development through mechanisms of concessionary aid and domestic borrowing than through foreign direct investment. As an indication, in 1977, Tanzania received approximately $300,000,000 in aid, equivalent to $20 per person. This per capita figure must be one of the highest in the world.

As a newly independent government in 1961, Tanzania inherited industrialization strategies based on reports and studies conducted by consulting firms and international organizations. Still placing heaviest emphasis on agricultural development, Tanzania undertook its first five year plan in 1964 and proposed an expanded industrialization program that was to be heavily reliant upon domestic and foreign investment from the private sector. Mechanisms to achieve this growth were to include accelerated depreciations, guarantees for repatriation of capital, and tariff protection. Market expansion was to take place by imposing constrictions on East African Common Market partners in order to protect local import substitution industries. In addition, a licensing procedure was proposed to reserve the common market for those national industries having comparative advantage over the other national partners.

In the midst of the first five year plan, President Nyerere reversed the industrial strategy based on foreign investment as the major instrument of development and set a new course as announced in the Arusha Declaration, 1967. It was announced that dependence on foreign investment would not be consistent with socialism, self-reliance, and political independence. The Arusha Declaration laid out two major goals: (1) the gradual substitution of private ownership of industry with state ownership—the creation of a system of parastatal corporations; and (2) the constraint and gradual replacement of private ownership and control of agricultural production by collective ownership and production within the framework of the Ujamaa village (based on the principle of cooperative activity among people and aimed at organizing the highly dispersed population in the potentially most productive areas of the land). A main policy accompanying the Arusha Declaration was the goal of increased employment in industry. The Arusha Declaration did not explicitly downplay investment in industrial production nor did it proscribe a restructuring of production. It transferred the means of production to the state.

Just how unclear Tanzania's post-Arusha Declaration industrial strategies were is evident in the publication of the second five year plan in 1969. It indicated that "an important task which will be undertaken early in the second plan period is the preparation of a long-term industrialization plan." Emphasis was placed, however, on labor intensive and small scale, local industries. The industrial section of the plan called for the production of simple import substitution goods, intermediate and capital goods, products for export, and simple manufactures from small-scale firms. The decentralization of industry was also to be accompanied by some sort of social cost-benefit analysis.

Something akin to a debate centered on what kind of strategy should be followed for long-term industrialization. The options that were considered fell within a socialist context and were discussed and then incorporated in the third five-year plan in 1974. The winning lobby advocated the development of basic industries, that is, those that would fulfill the primary needs of a poor society. Therefore, industrialization is aimed at import substitution and the processing of agricultural commodities. Foreign investment is still seen as detrimental to the goal of self-reliance and independence.

Scientific and Technological Infrastructure. The Tanzanian government is aware of the value of a scientific and technological infrastructure to support its national development in general and its productive capacity in particular. Organizationally, however, it is in a most elementary stage. The nationalization of key industries under the administration of parastatal organizations has served political ends but not productive ones. Overmanning, false starts, underutilization of existing plants, shortages of skilled and managerial talent—all have contributed to shortfalls in production.

Laboratories, technical institutes, management development institutes, and other facilitating arrangements are not capable of developing or adapting technologies to meet productive needs. Foreign sources, chiefly through mechanisms of foreign assistance, are valued as suppliers of industrial equipment and accompanying productive know-how. It is increasingly clear that to serve immediate production needs, training centers for specific industries are literally being attached to operating plants to solve manpower problems. These centers are shortcuts around the general educational system.

With little emphasis placed on foreign direct investment and licensing agreements, most technology is transferred by consultants from international organizations or traditional consulting firms. In addition, expatriates, mostly South Asians, are seen as sources of technology importation. They are, however, unpopular sources because of the relatively high salary levels they command. The connection is between these foreign sources on the one hand and the parastatals and their ministries on the other. Specialized institutions to act as decision makers and evaluators of technology importation have yet to be established on a viable basis. The Tanzania National Scientific Research Council is beginning to grapple with policy questions such as what Tanzania should do to utilize science and technology effectively. The real operational body that evaluates competitive bids among suppliers of technology and products is the Tanzania Investment Bank, which coordinates decision making with appropriate ministries and parastatals.

A poor country such as Tanzania is often in an uncontrollable position with regards to terms of trade, unilateral oil price increases, droughts negatively affecting production of cash crops, and similar problems. It is caught in a vicious cycle every few years. For Tanzania, capital requirements are met largely by foreign aid and loans at favorable terms. According to U.S. embassy sources in Dar es Salaam, development loans and grants for 1976-77 stood at $217,000,000. The pledge for 1977-78 rose to $268,000,000 of which more than half is in the form of grants. This suggests that Tanzania needs institutions that do not now exist and expertise (probably foreign) to concentrate primarily on technology selection. The goal of self-sufficiency is certainly long-term and dependent upon outside assistance. First generation technology generation appears to be years away for Tanzania.

Scientific and Technological Infrastructure

Personnel Sources. Roughly 85 percent of the population is illiterate. Only recently was education made compulsory, but there are obvious problems in terms of facilities, materials, and teachers. As late as 1964 there were only 630 secondary school teachers in the entire country and only twenty of these were Tanzanian. Plans call for education related to work. The first seven years consist of primary education and practically all succeeding years relate to training for work. There is one university in Tanzania, and one institution specializes in business administration. Foreign institutions predominantly in Europe and the United States are utilized for advanced management training.

Technology Sources for Individual Firms. Parastatal corporations engaged in industry tend to rely on foreign sources for capital goods as well as managerial assistance. The cement industry undergoing extensive expansion relies upon foreign financial sources for loans in order to purchase entire turnkey plants. Local suppliers are not available to produce parts of the industry specific technology. Balance of payments pressures create problems for necessary spare parts because of import restrictions. And training of personnel to man and maintain the plant is dependent upon foreign grants to make available foreign specialists to design and manage vocational centers. This example is typical of most large-scale industrial projects. It is also true of service industries that usually rely upon foreign consultants to evaluate and upgrade inventory and other control systems as well as to design and often provide the needed training. Expatriates and volunteers from private organizations such as the International Executive Service Corps provide both system-specific and firm-specific guidance. Traditional consulting firms such as Arthur D. Little, Inc., Cambridge, Mass., and international organizations such as the World Bank and the UNDP also provide technical assistance.

Technology Diffusion Networks. Except for a few professional associations of doctors and others, no networks exist for the diffusion of industry-related knowledge. In an area of critical shortage such as accounting, there are so few equivalents to CPAs that no professional association is feasible.

A noteworthy exception is the Small Industries Development Organization established in 1973 as a national parastatal organization. With offices in each region, SIDO is responsible for planning, coordinating, promoting, and offering almost every form of service and technical assistance to small-scale industries on a national level. An extension staff includes economists, industrial engineers, and other technologists in specific industries. In addition to consultancies, training both in formal settings and on the job, and technical services, the development of relevant technology for rural areas is facilitated through model demonstration projects and then passed on to relevant authorities. SIDO is primarily the only national organization of its type to diffuse small industries technologies and know-how throughout rural areas.

The nature of parastatals as monolithic industrial sectors makes the establishment of diffusion networks based on industrial groupings illogical. But networks and associations based on management functions—marketing, accounting, quality control, and inventory control—are sorely needed to spread what little expertise exists.

Political Climate for Business

Local Political Climate for Business. The Arusha Declaration in 1967 and the Mwongozo leadership guidelines of 1971 are ethical tracts laying out a revolution of attitudes that aim to produce a democratic society. The instruments to achieve Tanzania's hoped-for version of socialism are the parastatal means of production and the political parties, Tanganyika African National Union (TANU) and the Afro-Shirazi Party, which have been merged into the Revolutionary Party, fortifying the mainland and Zanzibar union. The political and economic considerations are never far removed from each other. At the top levels of parastatal management, national objectives are sought through economic means. And at the firm level, workers' councils have been organized to participate in decision making. At present, such participation is not effective because of low educational levels of the workers and the lack of clear policy direction at top levels. Nonetheless, the political party and the workers' councils are increasingly raising political consciousness and, barring any sudden crises, will increasingly limit the growth of the private sector.

Julius Nyerere has been president of Tanzania since independence in 1961. His policies are pursued aggressively, with impacts on business organizations and incentives. The relationships between urban and rural incomes has been a topic of interest in pursuing the national objective of equity. In 1961, the urban wage-salary relationship to rural income was seventy to one. By 1974, this was reduced to fifteen to one. It now is nine to one. The goal is to let everyone have access to something, but to prohibit luxuries for the few. In addition, government authorities take strong actions, with public support, against migration to the largest cities. Loafers are transported back to villages. When coupled with a decreasing productivity and underutilization of existing plants, a dilemma emerges: how to provide incentives to workers and still maintain a relatively equitable distribution of income.

Relations with the United States. The official U.S. relationship is confined largely to current assistance policies of meeting basic human needs. This relationship is expressed through loans which in 1975 totaled around $20,000,000, and through grants amounting to $7,000,000 in 1975.

Unofficially, there is a small overseas U.S. business presence because of market size. Instead, traditional consulting firms and private voluntary organizations operate well and freely at all levels of the government.

Economic Factors

At the time of independence in 1961, Tanganyika was one of the world's poorest countries. By U.N. standards, today it is still in that category. In 1961, roughly 30 percent of the economy was involved in subsistance production; by 1973, that number had dropped only a few percentage points. The economic picture indicates problems that stem from many factors indigenous to Tanzania and are not traceable only to errant strategies:

Table I-14. Tanzanian Economic Growth (1964-1976)
(in 1966 prices)

	1964	*1976*
GDP ($ millions)	680	1,196
GDP real growth rate (percent)	2.7	5.2
Share of Manufacturing in GDP (percent)	7.0	10.0
Share of Agriculture in GDP (percent)	47.0	38.0
Share of Administration in GDP (percent)	11.0	17.9
Gross Capital formations as percent GDP	13.0	22.0

Part of the problem with the slow growth rate (and bearing in mind a birth rate close to 3 percent) is that there has been a low level of investment in directly productive activities. The third five-year plan will attempt to remedy this situation by emphasizing the productive sector that will receive more than 40 percent of the investment budget.

Coupled with recent low levels of investment in production is reduction in output per worker in manufacturing and stagnation in output per agricultural worker. Related is the relatively poor performance of the parastatals, roughly 40 percent of which operated at losses. Endemic problems of overstaffing, lack of cost and accounting controls, and shortage of skilled and managerial talent are evident. In addition, many parastatals operate at far below capacity for internal managerial reasons and because of a lack of spare parts, physical infrastructural deficiencies, and maintenance and repair inadequacies.

The issue of interdependence (or dependence, depending on one's viewpoint) is acutely felt by the government. The 1973-1974 oil price increase, as one example, had a severe impact on the economy. In 1972, Tanzania imported a little more than 1,000 tons of petroleum at an approximate cost of $19 million; in 1975, less was imported at a cost of $50 million. In addition, world stagflation and the drought during 1973-1975 has placed Tanzania in a game of catch-up. These factors have contributed to decisions to impose import restrictions on consumer and other nonessential items, to improve productivity in export commodities, and to create import substitutes. The goal is to move the economy towards processing domestic raw materials for consumption in the local market.

Four agricultural products—coffee, cotton, sisal, and tobacco—accounted for more than 60 percent of export earnings in 1976. The high price of coffee on the world market presented Tanzania with a bonus in 1976. Coffee exports earned 31 percent of all export earnings, even though coffee production declined. This kind of fluctuation, while positive, can become a severe problem (the oil import costs, for example). It is the overall international market that has been controlling the Tanzanian economy, rather than Tanzanian officials finding a way to meet its own objectives within an international context. It is little wonder that the development of basic industries and improvement in agricultural productivity are seen as ways to reduce dependence on other countries.

Physical Infrastructure

Energy. Interruptions and shortages of power are problems for the industrial sectors. Tanzania has great hydroelectric potential and an undetermined and untapped amount of coal reserves. Electricity sales in 1976 were shown at 527 million kilowatt hour, up almost 9 percent over 1975. This rather large (for Tanzania) increase in demand has prompted the government to seriously consider to three power generation projects.

The rapid price rise in OPEC supplied crude oil had a detrimental effect on Tanzania's balance of trade, as noted. The impact in terms of inflation and trade balance forced the government to institute policies of trade restriction and limitation of foreign ownership by international oil firms whose main activity in Tanzania is crude processing. Tanzania has little in the way of oil reserves.

However, natural gas was found in significant quantities (thirty billion cubic litres) at Songo Songo, an island about twenty miles off the southern coast. While the Ministry for Water, Energy and Minerals claims that more than 80 percent of the deposit can be used for commercial purposes, it has not yet been determined what is the most efficient mix of uses for the reserves.

The OPEC price increase has also compelled Tanzania to investigate the feasibility of developing, with foreign expertise, renewable energy sources using appropriate technologies. Solar energy is one obvious consideration, given Tanzania's population dispersion and low-level but required use for village industries.

Transportation. Transportation is particularly important because of the wide dispersion of the population in rural areas. Tanzania has approximately 21,000 miles of roads of which only about 20 percent are paved. The poor condition of the roadway is one key reason for the high incidence of vehicle breakdown, inhibiting personal travel as well as distribution of goods.

There are about 2,800 miles of railroads including the 1,150 mile Tam Zam railroad completed with the assistance of the People's Republic of China. The shortage of ground transportation is somewhat offset by the internal air transport of Air Tanzania, though this is inadequate as a result of the dissolution of the East African Community and recurrent problems of mechanical maintenance.

Communication. The telephone system does not adequately link the entire country. Tanzania is planning to complete a microwave network to link the coastal grid with interior regions.

Although much could be done to improve the physical infrastructure, the government has decided, according to the third five-year plan, to put less development resources into this area and more into the directly productive sector. Between 1970 and 1974 economic infrastructure averaged almost 48 percent of the total government development expenditure. During that same period, less than 23 percent of government spending was channeled to the directly productive sector. By third five-year plan standards, these numbers were close to being reversed with 43 percent for the directly productive sector.

Cultural Patterns of Business Activity

Social Characteristics. Very high levels of illiteracy (85 percent) and infant mortality (150 of every 1,000 live births), a relatively short life expectancy (47 years), and the rural predominance of social organization continue to characterize Tanzania. Industrialization efforts often must bypass the educational system which is too new and small to generate manpower in necessary numbers and quality.

The Ujamaa village development policy is both rooted in the history of what is now Tanzania and aggressively implemented as part of policy. Village organizations existed in the past for a number of reasons. Chief among these is the fact that Tanzania has more than 130 ethnic groups, only one of which, the Sukuma, exceeds one million members. While tribes still do exist, there has been no dominant tribe that has attempted to divide the country politically. Non-Africans, numbering less than 1 percent of the total population, include Indians and Pakistanis, Arabs, and Europeans. While many Europeans still heavily influence Tanzania, nationalization has obviously significantly decreased their position and power in economic and political institutions. Indians remain an important factor in trade and merchandising, especially in the private sector.

There is an almost equal religious mix of Christian, Moslem, and animism. President Nyerere had a Christian upbringing and was educated in missionary schools and abroad. It is difficult to determine the importance of religious values in Tanzania's development, however.

The routes to power and influence in Tanzanian society are through the civil service establishment and through commercial farming, especially in crops and commodities for export. The parastatal managers have great access and power primarily because they are the few who have acquired knowledge and skills that are in very short supply. There is a tendency to rotate this elite into productive areas that are either trouble spots or tapped as a priority. The disruption such change usually causes within the ranks of subordinates testifies to the importance and need of top managers. In addition, many key middle level managers are expatriate Indians or Pakistanis whose presence is not always welcomed but whose expertise is badly needed.

Worker Attitudes. Although Tanzania is an agrarian society, the people are still strongly attracted to the cities. One reason is the difference in income between the city and the village, the former being nine times greater than the latter. However, there are few opportunities for employment in the cities, and it is not uncommon for the newspaper to report the roundup and return of urban loiterers to villages.

Education and specialized training are keys to future success in the Tanzanian money economy. Most parastatal heads have been educated abroad at English or American universities, and important middle level managers have had the benefit of advanced managerial training in institutions such as Arthur D. Little or the East African Management Development Institute, run primarily by U.N. affiliated organizations.

Overstaffing at lower, worker levels is a characteristic problem of parastatal corporations. This is the result of the government intervention through nationalization of the means of production. In other words, employment is an area of concern at least as much to politicians as to economists. This political dimension is reinforced through participation of TANU—the single political party on the mainland—whose purpose includes representing worker needs and concerns before management and raising the political consciousness of workers to truly participate in decision making at the firm level. This kind of benefit is a nonfinancial substitute for economic incentives that market systems provide to workers. However, it is difficult to criticize the lack of financial rewards because of the stated objectives of the government. Equity is one goal that will not be sacrificed; thus, efficiency of production is traded off for a political belief.

Productivity trends for Tanzanian manufacturing firms are to be suspected because of fairly unreliable data. The statistics that do exist in the Ministry of Finance and Planning indicate a decline in industrial growth despite a high level of investment. Between 1969 and 1974, for example, the incremental capital output ratio for manufacturing parastatals rose from 1.46 to 5.60. At the same time, the rate of return on capital during the same period declined from 23.0 to 12.6. In addition, for the same period, output per worker in parastatal manufacturing firms had also fallen from $871 to $636. It is interesting to note that comparable figures for private manufacturing firms are somewhat more positive. For the same period, output per worker increased from $643 to $757, while the rate of return on capital increased from 18.0 to 44.0.

There are numerous problems facing parastatal production in manufacturing. The most serious, in addition to overall balance of payments problems and the importation of inflation, is the trade-off between efficiency and equity. There are some indications that inefficiencies in parastatals will not be tolerated (layoffs of overstaffed firms, for example), but the real problem lies in the training, educating, and upgrading of technical and managerial skills. The dilemma of how to increase worker productivity without financial incentives must also be resolved within the framework of a socialist society.

Dominant Industries and Markets

Traditional Industries. At independence, Tanzania inherited an agriculturally based economy developed along plantation lines during the colonial period. Many of the family-run plantations still exist independent of state control and are responsible for a significant amount of exports. Agricultural production accounts for about 40 percent of GDP while manufacturing generates only 10 percent. Of the traditional agricultural production, increases during 1976 were registered in cotton (19 percent) cashewnuts (37 percent), and tobacco (35 percent), while decreases were experienced in coffee (9 percent), pyrethrum (21 percent) and sisal (9 percent).

Markets. The value of Tanzanian agriculture accounted for 67 percent of export earnings in 1976. Coffee, cotton, sisal, and tobacco alone earned about $300 million or two-thirds of all export earnings.

Manufactured exports in 1975 represented 15 percent of total exports and had a declining share dropping from 18 percent of total exports in 1971. In 1975, the most important manufactured products for export included sisal products (24 percent of total manufacturing exports); animal and vegetable oil products (13 percent); and cashew kernals (12 percent). In addition, petroleum products represented 5 percent of all exports. The decline in manufactured exports occurred as a result of market and political difficulties in the East African Community, lagging production of many agricultural raw materials, decline in productivity in general, and increased domestic demand for a number of exportable manufactures, such as meat and vegetable and animal oil products.

In 1976, all exports accounted for about 19 percent of GDP. The export of manufactured products was quite small, only about 2 percent of GDP.

New Industries and Markets. Limitations of natural resources and manpower necessarily have limited the government's planning for new industries. Emphasis therefore is being placed on traditional industries with an eye to creating or expanding industries with some potential. Since there is not much risk capital, mistakes are extremely costly, reverberating throughout the entire economy. To strengthen the internal economy and to deal with problems of maintenance, most recent planning and the budget emphasize meeting recurrent expenses rather than developmental expenditures. Of the recurrent budget (almost 60 percent of the total), advancements are hoped for in primary education, infrastructural improvements and bolstering of principal agricultural crops and food grains.

In the longer term, the government intends to achieve a self-sustaining economic base by prudently promoting industrial expansion and diversification in order to meet domestic demand, adding value to commodity exports, and creating import substitutes. The current, third five year plan has stated the goal of channeling 43 percent of the investment budget into the directly productive sector, accentuating the use of local inputs in the manufacture of goods for domestic consumption. The plan gives priority to the expansion of sugar plants, existing textile weaving facilities, and cement plants; the construction of a pulp and paper plant, a pharmaceutical operation and a food processing plant; and continuing study for developing coal, natural gas, and other natural resources.

This plan for a relatively high degree of self-sufficiency is so dependent on external factors that the integrity of the plan is certainly questionable. A shortage of skilled manpower, a dependence on external pricing structures, international financial assistance, expatriate expertise and foreign technology—all are areas of recognized importance at the parastatal, firm level. These areas of deficiency and dependence will condition the Tanzanians' ability to meet stated goals. As one example, the cement industry until very recently consisted of one plant operating below capacity and at low productivity rates. Before any ability was demonstrated to operate the single plant at full capacity and efficiently, a governmental decision was made to expand the number of plants to three. Although the cement industry expansion plan includes a heavy training component, it is generally typical of the government to invest in plant and equipment without commensurate concerns over productivity of labor.

Issues Affecting Firms

Business Relations with Government. At the parastatal level, business and government are essentially one and the same. There is a relatively short history of relationships between government and state-run enterprises that were instituted in 1967 in the Arusha Declaration. The parastatal system, while representing about one-half of industrial production, is very important to Tanzania's future. It is from these large corporations that the government hopes earnings will be generated to reinvest in further development. However, as mentioned earlier, a series of serious management and operating problems must first be solved. At the same time, there is the persistence of the private sector, which will probably diminish over time as greater amounts of investment are channeled into the parastatals. In addition, the large amounts of foreign assistance (approximately $300 million) are aimed at the parastatals as well as other parts of the government, thereby ensuring less attention to the private firms.

A key feature of parastatal/governmental relations is the excessive control mechanisms reflected in greater administrative expenditures.

Operating and policy decisions are taken in a hierarchical fashion through a chain including manager, board of directors, parent ministry, Ministry of Finance and Planning, Economic Committee of the Cabinet, and Parliament. This formal authoritative hierarchy is supplemented with a host of other controls over most aspects of investment and production decisions, such as the following: approval of annual investment plans, restricted use of the profits of a given parastatal, approvals to import both capital and goods, setting of wage scales and output prices, and others. These many different layers of regulation have made coordination of industrial planning difficult at best. The different functions of the control elements are often vague and redundant, leading to confusion over what actually constitutes policy.

Another important area of official policy is the establishment of small-scale industries in the village environment. The Small Industry Development Organization, itself a parastatal, has the mission to assist in all phases of small business development. Yet, the growth of large-scale parastatals threatens to severely limit the role that small firms will play in the economy by preempting the market.

Attitude Toward Foreign Technology. Tanzania does not have the institutions or managerial capability of developing or adopting foreign technologies to any significant degree. While there is recognition that both hard technologies and know-how are required from foreign sources, there is no institutional capability or responsibility to determine local needs for technology or search behavior. The approach is haphazard, and initiative seems to be taken by foreign missions in Tanzania or by international organizations such as U.N.-affiliated agencies or the World Bank.

Top management in parastatals, for example, knows fully well that Tanzania is dependent upon foreign suppliers for technologies—both appropriate and large-scale industrial types. Mid-level managerial and skilled manpower deficiencies inhibit adaptation and maintenance requirements. Thus, volunteer and paid consultants, U.N. and World Bank technical advisers, and especially expatriates are relied upon to fill technical and know-how voids. Where such external advisers are absent, maintenance of equipment, productivity, and efficiency of operations suffer. To remedy this problem, training facilities are being incorporated within many parastatals at the factory or plant site to teach fundamental skills so that simple repairs can be made. Until success is demonstrated at these levels, it is unlikely that adaptation or innovation can be managed successfully, even with the help of foreign experts.

In 1967, at the time of the Arusha Declaration, according to the Bureau of Statistics, there were sixteen parastatal corporations. By 1977, the government either owned directly or partially (at least 51 percent) 330 parastatals of which 175 were engaged in business. Of these business organizations, more than sixty suffered financial losses. The government did not create new corporations; it merely nationalized rather large, private firms with no accompanying increase in efficiency.

Some present parastatals are large by any standards, some employing up to 4,000 workers. The vast majority of value added in manufacturing enterprises occurs in enterprises of ten or more employees. Statistics clearly delineating small industry from large and, at the same time, private from public firms were not available to draw comparisons with the stated national policy of villagization.

The relationship between workers and management has already been mentioned. This relationship is undergirded by TANU, the political party, whose function is to insure worker participation in decision making. The Mwongozo guidelines of 1971 emphasized this new collaborative effort. Equity as a national goal among all Tanzanians is being achieved to some extent by wage and price guidelines that tend to minimize differences in wages and salaries. Much of the implementation of labor policies has received assistance from the International Labor Organization. The current ratio of urban to rural wages is nine to one.

The effects of the equity and worker participation aspects of labor relations practices have been the following: procedures for terminating worker employees are extremely cumbersome with workers' councils protecting laborers at the plant level; bonuses and pay scales to induce greater productivity are virtually nonexistent. Again, as mentioned previously, efficiency and productivity are traded off for equity. Nonetheless, where greater skills are required, a difference in wage is reflected. For example, in 1973, when GDP per capita was approximately $115 a year, workers in the consumer goods industry were averaging roughly $660, workers in intermediate goods production averaged $1,060, and the small number in capital goods production earned $1,160.

NOTES

1. Hyung Sup *Choi*, speech delivered at the East-West Center, Honolulu, February 1977.

2. Rubem Novaes, "Brazilian Investment Policy Priorities, Procedures and

Perspectives" (presented at the Brazil-U.S. Business Council meeting, Washington, D.C., October 17-20, 1977), p. 3.

3. Cláudio de Moura Castro, "Investmeno Em Educao No Brasil: Um Estudo Sócio-Econômico De Duas Comunidades Industriais" (Rio De Janeiro: IPEA/INPES, 1973), p. 13.

4. Novaes, p. 1.

5. Ibid., p. 2.

6. Ibid., p. 3.

7. Ney Braga, Minister of Education, comment at Superior War College, Rio de Janeiro, July 1976.

8. *National Development Plan II (1975-1979)*, (Brasilia: IBEG, Graphic Services, 1974), pp. 26-47.

9. Ibid., pp. 28-29.

10. *Government Programs in Support of Technology Development*, (Sao Paulo: State Council of Technology, October 1975), p. 4.

11. National Development Plan II, p. 102.

12. Ibid.

13. *Government Programs in Support of Technology Development* (Sao Paulo: State Council of Technology, October 1976), pp. 14-15.

14. Novaes, p. 18.

15. *Government Programs in Support of Technology Development*, 1975, pp. 19-21.

16. Denis A. Daniel, "Trademark Policy in Brazil Under the Second Basic Plan for Scientific and Technological Development," *The Trademark Reporter*, May-June 1976, p. 173.

17. James Bruce, "INPI—Controlling the Inflow of Technology," *Brazilian Business*, (Sao Paulo: Chamber of Commerce of Brazil, September 1976), p. 3.

18. *Education in Brazil: A Key Factor for Development* (Sao Paulo: American Chamber of Commerce for Brazil, 1977), p. 14.

19. Braga.

20. *Education in Brazil*, p. 6.

21. Observation drawn from Ramon Barquin, research on Brazil used as basis of Chapter 7 in *National Control of Foreign Business Entry*, Richard D. Robinson, ed. (New York: Praeger, 1976), p. 202.

22. Novaes, p. 4.

23. Small firms are defined as those employing between one and one hundred persons, medium firms as those employing between one hundred and five hundred employees, and large firms as those with a staff of over five hundred persons. For analytical purposes, technological change is defined as 1) the introduction of a new product; 2) the modification of an existing product or product line; 3) the utilization of a new manufacturing or preparatory process; and, 40 modifications in production and/or administration methods to improve quality and production efficiency.

24. Barquin.

25. Braga.

26. Monograph, "Technological Change Within the Firm—Factors Influencing Behavior of the Entrepreneur" (Sao Paulo: Sao Paulo State Council of Technology, October 1975), p. 8.

27. Richard L. Seely. "A Working Paper on the Transfer of Managerial Technology to Developing Countries: The Case of Brazil" (1977), p. 68.

28. Eugene McCann, "An Aspect of Management Philosophy in the United States and Latin America," *Academy of Management Journal*, June 1964, pp. 149-152.

29. Barquin.

30. Terrence Cullinan, "Latin American Education and Recruitment: An Environmental Perspective." *California Management Review* 12, III, (Spring 1970), p. 43.

31. Seely.

32. Washington: Embassy of Brazil, No. 8, June 15, 1973, p. 1.

33. Monograph, "Control System for the International Transfer of Technology and Capital to Brazil," (Sao Paulo: Sao Paulo State Council of Technology, October 1975), p. 4.

34. Novaes, p. 5.

35. P.G. Garland, "Doing Business in Brazil" (Rio de Janeiro: Bolsa de Valores de Rio de Janeiro, 1973), pp. 66-67.

36. *National Development Plan II*, pp. 26-38.

37. Daniel, pp. 173-82.

38. Kerry Fraser, "Education in Brazil: A Key Factor for Development" (Sao Paulo: American Chamber of Commerce for Brazil, 1979), p. 19.

39. Novaes, p. 19.

40. Ibid., p. 18.

41. Monograph, "Control System for the International Transfer of Technology and Capital to Brazil," (Sao Paulo: Sao Paulo State Council of Technology, October 1975), p. 15.

42. Monograph, "Technological Change Within the Firm—Factors Influencing Behavior of the Entrepreneur" (Sao Paulo: Sao Paulo State Council of Technology, October 1975), p. 8.

43. Fraser, p. 26.

44. Same monograph as footnote 41.

45. Fraser, p. 6.

46. Braga.

47. Fraser, p. 6.

48. This according to the Brazilian Center for Management Assistance for the Small and Medium Company (CEBRAE).

49. Same monograph as footnote 41, p. 1.

50. Novaes, p. 14.

51. Ibid.

52. James Bruce, "INPI—Controlling the Inflow of the Technology," *Brazilian Business*, (Sao Paulo: Chamber of Commerce of Brazil, September 1976), p. 4.

53. Same monograph as footnote 42, p. 7.

54. "Editorial," *Industria Peruana*, Lima, No. 524, November-December 1976.

55. *Your Business Partner in Africa: An Investment Guide*. Nairobi: Ministry of Commerce and Industry and Kenya External Trade Authority, 1977, p. 12.

56. Kibaki's Plan to Make 'All' Kenyans Rich," *Weekly Review*. Nairobi, December 12, 1977. p. 10.

57. *Economic Survey 1977.* Nairobi: Central Bureau of Statistics, 1977, p. 10.

58. *Kenyanization of Personnel in the Private Sector: A Statement on Government Policy Relating to the Employment of Non-Citizens in Kenya.* Nairobi: Ministry of Labour, 1967.

59. *Economic Survey*, p. 157.

60. Ibid., p. 157.

61. Michael T. Kaufman, "An African Dilemma: Using Foreign Experts," *International Economic Survey*, *New York Times*, February 5, 1978, p. 92.

Appendix II: Description of Research Program

INTRODUCTION

The research program on the transfer of managerial technologies attempted to identify what kinds of factors affect the transfer of technology and problem solution resulting from IESC volunteer consultations. This statement of purpose led to the development of an interactive model which identified internal firm and external factors (independent variables) that influence the process of consultation (independent variables) and the ultimate result of the consultation in terms of problem solution and transfer of technology (dependent variables). Several instruments were designed and employed in field research to collect data on these variables. The broad scope of the research and the difficulties of collecting comparable data in various countries argued for the development of detailed cases as the most useful and reliable form of data collection. The field researcher used to develop these cases was ultimately considered the most effective evaluator of the overall transfer process.

Instrument Design

As a result of an initial three month study of the International Executive Service Corps and its operations, decisions were made to design a field research program which would evaluate individual firms and how their behavior was changed through an IESC consultation program. It was felt in addition that research should attempt to describe generally the external environment with regard to its effect on local business enterprises. The instruments were designed to collect data on five general categories of independent variables and two specific dependent variables:

Independent Variables
- Firm management practices;
- Firm historical characteristics and resources;
- Diagnosis and process of consultation;
- External environment;
- Type of Technology.

Dependent Variables
- Problem solution;
- Degree of transfer of technology.

Previous research provided typologies that were used in developing the specific instruments. The process of management categories were drawn from Newman and Warren[1] and the major comparative management typologies came from work done by Neghandi and Prasad.[2] Several sources were used for information on change strategies and targets[3] and stages of a consultation program.[4] Concepts of turbulent environments came from Terreberry.[5] Several authors provided structures for the type of technology questions.[6] These focused on categorizing technology in terms of knowledge, stages of project development, and functional characteristics.

Data were collected on two target populations in the foreign environment. A field researcher visited individual firms in one of the five countries. He interviewed firm personnel following a general interview guide. He asked the interviewees (IESC clients) to complete a questionnaire, which was also completed by the IESC volunteer who worked with the client. The field researcher also evaluated the firm in terms of basic classifications of size, type of industry, and market domination. After all the case studies had been written, each researcher evaluated his cases in terms of whether or not technology was transferred as a result of the IESC project. This evaluation was conducted according to a different instrument, the field researcher evaluation, that provided another measure of critical dependent and independent variables.

In summary, several instruments were designed to collect data on projects from various viewpoints and to generate various kinds of information. The following list indicates the basic objective and content of all the instruments employed in the data collection:

- Questionnaire—five-point scaler attitude and evaluation questions for volunteers and clients;
- Interview guide and classification sheet—field researcher guide for interview and opportunity to obtain general classifying information such as size, management structure, and technology history;
- Field researcher evaluation—forty-two five-point scaler questions for each case answered by the field researcher after he had completed all of his cases;
- Case writing outline—the writing format used to assure common structure and coverage of all cases.

Instrument Tests

Three different types of tests were conducted prior to the field research to assure the effectiveness of the instruments for data collection. First, the questionnaire was submitted to a group of IESC volunteers to determine how long it took to complete and what areas were confusing or easily misinterpreted. Second, a field test in Peru of the Spanish translation questionnaire was conducted, again to test timing, clarity, and interpretability of the instrument. Third, a reliability test was conducted to determine whether or not volunteers, clients, and field researchers all answered questions about a single phenomenon in a similar fashion. A reliability test using the alpha coefficient was administered. This test yielded a reliability of .01 among field researchers and .06 among field researchers, volunteers, and clients. This was considered adequate for continued use of the instruments.

Field Data Collection

Following the field tests, revised instruments were used to collect data in five countries—Peru, Brazil, Kenya, Korea, and Tanzania. In each country, an environmental analysis was written; these are included in Appendix I. These pieces analyze the political and economic environments as they affect local enterprises. The field collection overseas generated the information on sixty-seven projects, based on the classification guide, and was the basis for the sixty-seven field researcher evaluations.

Phone interviews were conducted with those volunteers available who had worked on the sixty-seven projects examined, and they completed the same questionnaire that had been given to the client. This procedure provided the matching data from the domestic environment. To generalize findings from the study and to add information from different countries and projects, a survey of another 405 volunteers was conducted. The U.S. data thus consist of questionnaire information from sixty-seven volunteers who worked on projects examined by the field researchers and an additional 405 volunteer responses concerning other projects. The total data collected include:

Sixty-seven detailed cases;
Sixty-seven field researcher evaluations;
Seventy-one client questionnaires;
Four hundred seventy-two volunteer questionnaires (including the sixty-seven matching questionnaires); and
Five country environment analyses.

Analysis and Report Writing

Once the data had been collected, several inferential statistical techniques were used to analyze the data. These included regression analysis, canonical correlation, analysis of variance, cross tabulations, and factor analysis. The research group and data consultants felt that the field researcher evaluations represented the most accurate and consistent responses. Through direct interviews and a review of the questionnaires, it was noted that in certain cases, terms were not fully understood by the clients and volunteers and produced some confusion in responses. It was felt as well that the field researcher had the greatest insight into the projects he analyzed and the least bias in terms of evaluating the project in terms of success or failure of technology transfer.

Figure II-1 illustrates the research tasks, the associated output, and the schedule for the research. Next comes a collection of the instruments used in data collection, which include the questionnaire, classification guide, field researcher evaluation, and the case study guide. Table II-1 displays selected sources consulted in designing the instruments and the use of the instruments.

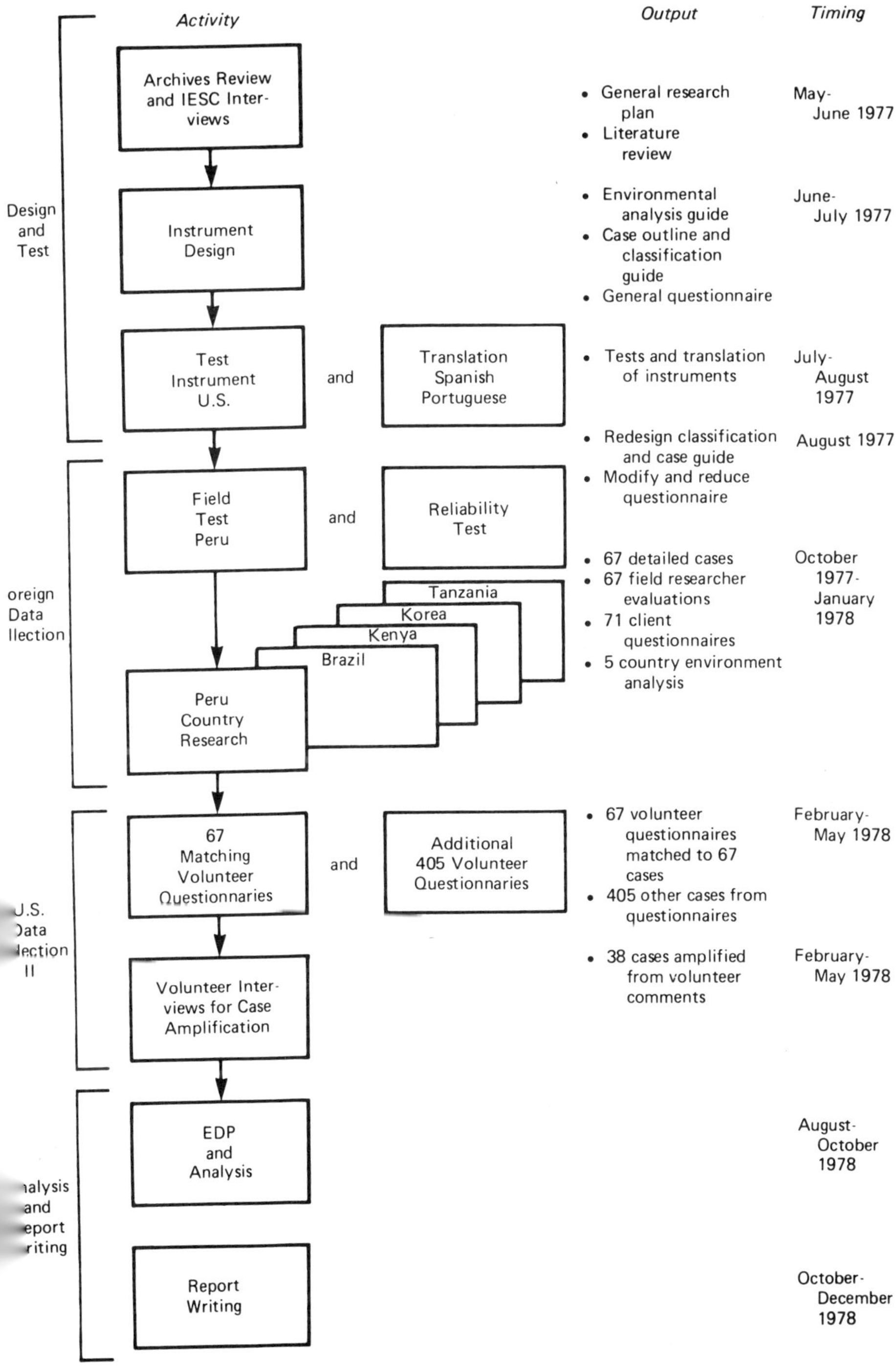

Figure II-1. Research Program

Table II-1. Selected Instrument Development Sources and Uses

Questionnaire, Classification Guide, and Field Researcher Evaluation			
Issue for Analysis	*Previous Research*[a]	*Example Characteristics*	*Instrument Used*
Management			
Process of management	Newman and Warren	planning, controlling, organizing, leading	classification guide, questionnaire
International comparative management	Neghandi and Prasad	decision-making structure, types of controls, motivation systems	classification guide
Functional descriptions	–	manufacturing, engineering	classification guide
Environmental impact analysis	Terreberry	turbulent, stable	classification guide, field researcher evaluation
Process of Consultation			
Change strategy	Hornstein et al.	data based change program	classification guide, questionnaire
	Blake and Mouton	acceptance, confrontation, catalyzing	classification guide, questionnaire
Process of consultation	Ginsberg	diagnosis, implementation, termination	classification guide, questionnaire, field researcher evaluation
Target of change	Tichy and Hornstein	senior level, middle management	classification guide, questionnaire
Type of Technology			
Knowledge categories	Svennilson and Peno	firm specific, system specific	classification guide, questionnaire, field researcher evaluation
Project development stage	Behrman and Wallender	design, construction, value engineering	classification guide, questionnaire, field researcher evaluation

Situation	–	crisis, new market opportunity	classification guide, questionnaire, field researcher evaluation
Attributes	U.S. Study of Agricultural Extension Institutions	money saving, time saving, requires capital	questionnaire

[a]See bibliography for complete citation.

QUESTIONNAIRE

FMME
Fund for Multinational Management Education
684 Park Avenue, New York, N.Y. 10021
Telephone (212) 535-9386

Volunteer Executive

____

Project Number

____

Client Firm

____

Country

____

Project Dates

____

The Transfer of Managerial Technology to Developing Countries

The purpose of this study on the transfer of managerial technology is to generate information on those factors which condition the learning of new managerial skills by developing country businesses through working with volunteer executives like you.

The study draws on well established theories of management which describe techniques relating to all aspects of business (e.g., executive management, financial, production/operations, marketing/distribution, human resources). Executives like you represent vast experience in working with managers in techniques pertaining to these functional areas of management. Specific objectives within the study are to determine the degree and nature of the problem solution resulting from the IESC projects; to determine the extent and impact of technology transfer and technology diffusion within the client firm; and to isolate and analyze the critical variables which condition the success of problem solution, transfer of technology, and diffusion of technology which result from an IESC consultation.

Importance of Your Cooperation

Surveys of this kind depend heavily on a high response rate from the volunteers, and are important and useful when a large number of respondents cooperate. We would like to ask for your cooperation in completing the enclosed questionnaire. We do, of course, guarantee anonymity and confidentiality of responses. When compiled, the survey results will illuminate patterns reflecting conditions in the client firms or in the local environments which will be important indicators of what leads to successful transfer of technology.

Aids to Completing Questionnaire

The questionnaire should be completed with reference to one project only—if you have been on several IESC assignments, we ask that you please select one of these and answer all questions in the context of that project alone. Prior testing of the questionnaire indicates that it takes between one-half hour and one hour to complete.

The questionnaire was designed to be answered without the help of project files or any review work by you. In preparing questionnaires of this sort, it is difficult to word questions so that all seem relevant to the diverse projects which IESC has undertaken. However, we ask that you please make your best guess and try to complete all questions.

You will note that we use different scales for different sets of questions. Please refer to the scale provided for each group of questions.

Some questions are stated positively, while others are stated negatively. This variation is in accord with standard procedures for preparing questionnaires. Please read each question carefully.

Please note that the word "project" refers to your IESC assignment and project, and not to broader development projects of the client firm.

For each question, circle the number which corresponds to the answer you select. Please refer to sample below.

	Disagree strongly	Disagree	Neither agree nor disagree	Agree	Agree strongly
1 Industrial technology is valuable for developing countries.	1	2	3	4	(5)

The question as it is answered indicates that the respondent feels very strongly that industrial technology is valuable to developing countries. If, for instance, the number "2" had been circled, this would indicate that the respondent does not feel that industrial technology is valuable to developing countries.

We thank you for your willingness to participate in this study.

To what degree are the following traits important for an IESC volunteer like you to work well with the foreign client?	Not at all	Little	Neither much nor little	Some	Very much
1 experience in international business	1	2	3	4	5
2 prior consulting experience	1	2	3	4	5
3 specialized skills (skills applicable to one industry only)	1	2	3	4	5
4 generalized skills (fundamental management principles applicable to many industries)	1	2	3	4	5
5 familiarity with country	1	2	3	4	5

Here are ways in which an IESC volunteer can view his role:

as a ***problem solver*** who, when faced with a problem, can come up with a solution;

as a ***problem identifier*** who, when analyzing a management technique, can recognize what problems there are with it;

as a ***planner*** who, when analyzing a management technique, can devise ways of using most efficiently the client's resources;

as a ***teacher*** who, when dealing with the client, is most concerned with the new skills that the client is learning from the project.

Please refer to the scale shown at right for answering Questions 6-9.

To what degree do you view yourself as:	Not at all	Little	Neither much nor little	Some	Very much
6 problem solver	1	2	3	4	5
7 problem identifier	1	2	3	4	5
8 planner	1	2	3	4	5
9 teacher	1	2	3	4	5

The following management approaches were useful in the IESC project:	Not at all	Little	Neither much nor little	Some	Very much	Don't know
10 human relations	1	2	3	4	5	6
11 organizational behavior modifications	1	2	3	4	5	6
12 changes in information use	1	2	3	4	5	6
13 changes in structure	1	2	3	4	5	6
14 changes in tasks	1	2	3	4	5	6
15 introducing new system	1	2	3	4	5	6

My approach to the client can be categorized in the following manner:	Disagree strongly	Disagree	Neither agree nor disagree	Agree	Agree strongly
16 helping accept situation	1	2	3	4	5
17 catalyzing new efforts	1	2	3	4	5
18 confronting problems	1	2	3	4	5
19 prescribing solutions	1	2	3	4	5
20 providing theories or models to guide management	1	2	3	4	5

Please choose the answer which most closely describes the way your IESC project progressed.	Disagree strongly	Disagree	Neither agree nor disagree	Agree	Agree strongly
21 The IESC Country Director did a good job in identifying characteristics of the IESC project to which I was assigned	1	2	3	4	5
22 There was no problem in deciding how leadership would be shared during the IESC project.	1	2	3	4	5

	Disagree strongly	Disagree	Neither agree nor disagree	Agree	Agree strongly
23 The client's expectations for the results of the IESC project were unrealistic vis-a-vis the resources available to him.	1	2	3	4	5
24 The client did not tell me exactly what he wanted from the IESC project.	1	2	3	4	5
25 A large part of our time was spent with the client pinning down what he felt was his firm's problem.	1	2	3	4	5
26 Client personnel relevant to the project were just too busy to participate in the IESC project.	1	2	3	4	5
27 Having people other than the volunteer and client involved in the problem analysis was not felt by the client to be important.	1	2	3	4	5
28 During the IESC project, fruitful ongoing discussions were held with the client.	1	2	3	4	5
29 We never really reached agreement on what we expected from the IESC project.	1	2	3	4	5
30 During implementation, the client and I did not periodically rethink our goals.	1	2	3	4	5
31 As we implemented the system, we knew that some users were unhappy with how we were progressing.	1	2	3	4	5
32 Efforts were made to resolve the disagreements regarding how we were proceeding on the IESC project.	1	2	3	4	5
33 During the IESC project, I worked on a number of potential problems in addition to the specific problem.	1	2	3	4	5
34 Keeping our recommendations actively used will require continued outside support for the client.	1	2	3	4	5
35 In evaluating the system, the client was interested in criteria different from those we considered critical.	1	2	3	4	5
36 Few of the client's personnel relevant to the IESC project had learned to deal with management techniques proposed during the project by the time the project was completed.	1	2	3	4	5
37 The client will not have difficulty continuing to use the techniques introduced during the IESC project.	1	2	3	4	5
38 It was important that I maintain contact by mail to respond to the client's questions.	1	2	3	4	5
39 I did not stay involved with the client the right amount of time for this IESC project.	1	2	3	4	5
40 My IESC project changed the firm's product or service.	1	2	3	4	5
41 My IESC project changed the firm's production process.	1	2	3	4	5
42 My IESC project changed the organizational structure of the firm.	1	2	3	4	5
43 My IESC project altered the values of the client firm.	1	2	3	4	5
44 My IESC project altered how personnel are motivated in the firm.	1	2	3	4	5
45 The IESC project recommendations addressed the firm's operation at the top level.	1	2	3	4	5
46 The middle level of the firm's structure was the target of the IESC project recommendations.	1	2	3	4	5
47 The IESC project recommendations addressed the bottom of the firm's structure.	1	2	3	4	5

Please refer to the scale shown at right for Questions 48-52.

The client's function in the firm can be characterized as:	Disagree strongly	Disagree	Neither agree nor disagree	Agree	Agree strongly
48 entrepreneur	1	2	3	4	5
49 corporate management	1	2	3	4	5
50 linkpin (one who relates different departments of firm or communicates within entire organization)	1	2	3	4	5
51 task manager	1	2	3	4	5
52 bureaucrat	1	2	3	4	5

Please refer to the indicated scale for answering Questions 53-60.	Disagree strongly	Disagree	Neither agree nor disagree	Agree	Agree strongly
53 Some of the client's management personnel were never convinced that this IESC project was necessary.	1	2	3	4	5
54 It is not clear who in the client organization should be responsible for the implementation of the IESC project recommendations.	1	2	3	4	5
55 The client was not well prepared to begin work on the IESC project.	1	2	3	4	5
56 The personnel I conferred with did not seem to understand the problems the firm faced.	1	2	3	4	5
57 The management of the firm encouraged input from all personnel on matters concerning the firm.	1	2	3	4	5
58 Few of the client's management personnel indicated an interest in the recommendations I made for altering the system addressed during the IESC project.	1	2	3	4	5
59 The management personnel were not adequately prepared to utilize my suggestions.	1	2	3	4	5
60 New concepts were not discussed generally among management personnel.	1	2	3	4	5

Please note that for Questions 61-65, we are using a different scale. We would like to know how often the client discusses management techniques with others. You may answer "Not Applicable."

Within the firm, the client discussed management techniques with:	Never	Seldom	Occasionally	Frequently	Very frequently	Not applicable
61 president or chief officer	1	2	3	4	5	6
62 immediate superior	1	2	3	4	5	6
63 counterparts in other departments	1	2	3	4	5	6
64 subordinates	1	2	3	4	5	6
65 special committees	1	2	3	4	5	6

Please refer to the scale shown at right for Questions 66-69.	Disagree strongly	Disagree	Neither agree nor disagree	Agree	Agree strongly
66 There is an adequate coordination and communication mechanism for integrating the various departments in the client firm.	1	2	3	4	5
67 Management techniques of the firm's management personnel can be characterized as very sophisticated (e.g., use of computer information systems, complex planning for the firm).	1	2	3	4	5
68 The firm's management relies on routine accounting and cost data for its information on management policies.	1	2	3	4	5
69 The client firm's management processes rely primarily on personal observation and intuition.	1	2	3	4	5

Please refer to the scale shown at right for Questions 70-71.

	Disagree strongly	Disagree	Neither agree nor disagree	Agree	Agree strongly
70 Resources needed for implementing new management techniques are not available in the local environment.	1	2	3	4	5
71 The client is able to locate in the local environment skilled personnel for work in his firm.	1	2	3	4	5

Please refer to the scale shown at right for Questions 72-75.

	Disagree strongly	Disagree	Neither agree nor disagree	Agree	Agree strongly	Don't know
72 Government regulations encourage the client in applying new management techniques.	1	2	3	4	5	6
73 Government policies encourage use by the client of new management techniques.	1	2	3	4	5	6
74 Government programs facilitate the development of management personnel.	1	2	3	4	5	6
75 Local formal educational activities encourage training of management personnel.	1	2	3	4	5	6

Please refer to the indicated scale for answering Questions 76-84.

Outside the firm, the client discussed management issues with:

	Never	Seldom	Occasionally	Frequently	Very frequently	Don't know
76 local professional associations	1	2	3	4	5	6
77 local consulting firms	1	2	3	4	5	6
78 management training organizations (such as AMA)	1	2	3	4	5	6
79 university personnel	1	2	3	4	5	6
80 government agencies	1	2	3	4	5	6
81 financial institutions	1	2	3	4	5	6
82 managers in competing firms	1	2	3	4	5	6
83 foreign consultants	1	2	3	4	5	6
84 foreign firms	1	2	3	4	5	6

The term "technology" relates to the control and dissemination of knowledge. Knowledge which is used in the productive process can be categorized as:

General Business Knowledge:	that which is publicly available within the society, (e.g., through books, universities, etc.)
Industry Specific Knowledge:	that which is necessary to produce a product or manage a process and which is generally known within the industry (e.g., how to produce and market pharmaceutical products)
System Specific Knowledge:	that knowledge and know-how necessary for the production of a specific product or service (e.g., antibiotics)
Firm Specific Knowledge:	production process knowledge or know-how owned by or contained within a specific firm (e.g., the Xerox process)
Ongoing Problem-Solving Capability:	know-how resulting from experience and necessary to identify and respond effectively to general management principles

Please refer to the scale shown at right for the following questions.

In the project you have completed as an IESC volunteer, the knowledge you transferred can be described in which of the following ways?

	Disagree strongly	Disagree	Neither agree nor disagree	Agree	Agree strongly
85 general business knowledge	1	2	3	4	5
86 industry specific knowledge	1	2	3	4	5
87 system specific knowledge	1	2	3	4	5
88 firm specific knowledge	1	2	3	4	5
89 ongoing problem-solving capability	1	2	3	4	5

Attributes which characterize the IESC project are as follows:	Disagree strongly	Disagree	Neither agree nor disagree	Agree	Agree strongly
90 requires continuing expenditures to use ideas and suggestions	1	2	3	4	5
91 clear-cut pay-off for the firm	1	2	3	4	5
92 local outsiders feel the project was unnecessary	1	2	3	4	5
93 project will help client save time	1	2	3	4	5
94 project does not help client save money	1	2	3	4	5
95 project reduces physical discomfort for client	1	2	3	4	5
96 bottlenecks were removed	1	2	3	4	5
97 using the new ideas is risky for client	1	2	3	4	5
98 the management ideas from the project are compatible with client operations at present	1	2	3	4	5
99 the management concepts from the project do not require major commitment from the firm	1	2	3	4	5
100 the client firm can easily return to its previous management concepts after the project is completed	1	2	3	4	5
101 the project opened the client firm to no new types of innovations	1	2	3	4	5
102 many different members of client management personnel will have to participate to utilize the new technology	1	2	3	4	5
103 the project addressed technology concerned with organizing the firm's structure	1	2	3	4	5
104 the project aided in planning the client firm's activities	1	2	3	4	5
105 the project focused on controlling the firm's operations	1	2	3	4	5
106 the project assisted firm management in activating firm policies	1	2	3	4	5

For Questions 107-111, please answer according to the scale shown at right:	Disagree strongly	Disagree	Neither agree nor disagree	Agree	Agree strongly
107 The client never had much confidence in the amount of help that new management techniques could bring.	1	2	3	4	5
108 The client insisted on testing out the proposed solution prior to implementation.	1	2	3	4	5
109 The client seemed to feel that it was entirely my responsibility to solve the firm's problem for which I was contracted.	1	2	3	4	5
110 The client was interested in using new management techniques in his firm.	1	2	3	4	5
111 The client was not willing to pour whatever resources were required into IESC project.	1	2	3	4	5
112 The client was not interested in discussing with others in the local environment the techniques he learned from me.	1	2	3	4	5
113 The client was not anxious for his staff to learn new management techniques in the process of solving his firm's problems.	1	2	3	4	5
114 I believe that most foreign managers are reluctant to use U.S. managerial principles in their operations.	1	2	3	4	5
115 I do not believe that many foreign managers are capable of adapting U.S. managerial techniques for use in their firms.	1	2	3	4	5
116 I believe that one of the most important jobs in working with a foreign client is to change his general orientation toward management.	1	2	3	4	5
117 I could not productively have remained longer with the client on this IESC project.	1	2	3	4	5
118 I felt responsible for the IESC project's results.	1	2	3	4	5
119 The client is responsible for the results of the IESC project.	1	2	3	4	5

For Questions 120-132, please choose the answer which most closely reflects your opinion from the scale shown at right.	Disagree strongly	Disagree	Neither agree nor disagree	Agree	Agree strongly
120 The benefits we expected for the user never really materialized from the IESC project.	1	2	3	4	5
121 The IESC project provided the client with the assistance he needed.	1	2	3	4	5
122 The IESC project did not provide the client with techniques that helped him solve his firm's problems.	1	2	3	4	5
123 The IESC project did not provide the client with techniques that are valuable in solving management problems in his company.	1	2	3	4	5
124 The client tends to rely more on new management techniques in his work now that the IESC project has been completed.	1	2	3	4	5
125 As a result of the IESC project, decision-making and problem-solving procedures of the client have changed.	1	2	3	4	5
126 As a result of learning new concepts introduced during the IESC project, the client is searching for more new management techniques.	1	2	3	4	5
127 As a result of the IESC project, the client is not anxious to search for new management techniques.	1	2	3	4	5
128 The client is sharing the new management concepts with others outside his firm.	1	2	3	4	5
129 The client is not training others in his company in the techniques used in the IESC project.	1	2	3	4	5
130 Techniques involved in the IESC project were diffused to areas of the client firm other than the principal problem area.	1	2	3	4	5
131 The client is not continuing to use the techniques introduced during the IESC project.	1	2	3	4	5
132 Overall, I am quite happy with the results of the IESC project for the client.	1	2	3	4	5

PROJECT CLASSIFICATION GUIDE

I. Identification

1. Is project target a unit, subsidiary, or otherwise connected to other systems? (yes; no)

2. Industry (manufacturing; service; extractive; agribusiness; other)

3. Ownership of firm (government owned; privately owned, closely held; privately owned, many owners; joint venture with foreign participation)

4. Principal products

5. Number of employees (less than 50; 50-100; 100-500; more than 500)

6. Number of expatriates

7. Size of management group is (1-5 individuals; 6-15; 16-30; more than 30)

8. Sales of company (less than $100,000; $1000,000-$500,000; $500,000-$2 million; more than $2 million)

9. Sales growth of company, (over the three years preceding project, adjusted to 1977 index) (0-5 percent; 5-15 percent; more than 15 percent)

10. Employee growth of company, (over the two years preceding project, in addition to normal turnover) (no new employees; 1-10 percent increase; more than 10 percent increase)

11. Company is (in business 1-3 years; in business 3-6 years; in business more than 6 years)

12. Type of task technology (craft; routine; nonroutine; process)

13. Market share in principal product

II. Management Characteristics

Planning

14. Internal planning reflects (no planning at all; decisions made with no information generated specifically for situation; infor-

mal response to situations using cost data and routine accounting; informal planning and some group thinking concerning direction of firm; formal planning, some explicit quality control mechanisms; computer information systems, complex planning for organization)

15. Firm has (systematic and detailed plans; systematic but limited plans; randomly formulated forecasts and plans; no formal plans)

16. Frequency of review of plans (monthly; quarterly; semiannually; annually; ad hoc/as needed)

17. Planning covers (general environment developments; industry environment developments; firm environment developments; contingencies)

18. Temporal scope of planning is (random, no specific time; less than 1 year; 1-3 years; 3-5 years; more than 5 years)

19. Client uses these sources for information on management techniques (newspaper ads; technical journals, professional periodicals; equipment salesmen; trade fairs; contacts with competitors; manuals or textbooks; direct promotional mailing; visits to foreign firms; visits to other firms; government offices; contact with local consultants; contact with foreign consultants; foreign missions)

Organizing

20. Firm policies are (formally stated and explained to personnel; formally stated but not necessarily explained to personnel; not formally stated but generally known to personnel; not formally stated and not generally known to personnel; nonexistent)

21. General firm policies are formulated by (chief executive; chief executive and committee; special committee; other)

22. Acquisition of technology is inhibited by (lack of information on availability; technology available but does not fit firm capabilities; market for firm products is not large enough to support new technology; lack of capital or sources of financing; availability of raw materials necessary for technology not sufficient; government restrictions; lack of available servicing of technology; lack of management or skilled labor within firm to use technology; negative impact on labor; availability)

23. Management personnel selection is (formalized; informal)

24. Management personnel selections are made by (chief executive officer; chief executive officer and other managers; committee; other managers)

25. Training program involves (on the job training and in company formal training; on the job training and outside company formal training; on the job training only; formal outside training; ad hoc/as needed)

26. Promotion decisions are made by (chief executive officer; chief executive officer and special committee; special committee; other)

27. Layers of hierarchy (1-3; 4-6; 7-10)

28. Authority of definition via (charts; oral means; no formal means)

29. Authority communicated by (chief executive officer; staff; combination)

Leading

30. Major policy decisions made by (chief executive officer; top management and chief executive officer; top management, chief executive officer and relevant staff)

31. Production policies made by (chief executive officer; top management and chief executive officer; chief executive officer, top management, and relevant staff)

32. Marketing policies formulated by (chief executive officer; top management and chief executive officer; chief executive officer, top management, and relevant staff)

33. Strategic decision-making structure of company (dominated by individual or clique primarily; hierarchical; group participation in decision making; family dominance; political dominance; other)

34. Management trust in subordinates (subordinates viewed with confidence by management; condescending confidence demon-

strated by management; management demonstrates little or no confidence in subordinates)

35. Employee motivation linked to (monetary gain; monetary and non-monetary gain; organizational loyalty and monetary gain)

36. Employee participation in morale programs (all employees participate; employees concerns filtered up through staff; suggestion system available and used by all; staff filtering/suggestion system combination; suggestion system exists but is not effective; no formal program)

Controlling

37. Standard and procedure setting in production by (top management only; top management and relevant staff; senior staff member; no formal standards)

38. Standard and procedure setting in office by (top management only; top management and relevant staff (personnel); office manager only; senior staff member; no formal standards)

39. Scope of quality control (separate department for quality control; subsumed under production department; no formal quality control mechanism)

40. Cost control for (all products; certain products; no formal cost controls)

41. Budgetary controls (for entire company by senior management; for entire company by special staff; no formal budgetary control)

III. Process of Consultation

42. User department

43. Project was designed for (crisis situation; establishing new activity; modifying existing activity; instrumental for other change activity; defensive/rationalization; other)

44. Project request resulted from (response to market condition; response to unique technological capabilities within firm; response to government incentive; response to government disincentives; other)

45. Recommendations and techniques to be used by (single individual; two to five individuals; more than five individuals)

46. Number of people involved during diagnosis

47. Number of people involved during entry/planning

48. Number of people involved during implementation

49. Number of people involved during ongoing review

50. Project is primarily designed to change (product or service; production process; organizational structure; value of policies; people innovation)

51. Strategy for change (acceptance; catalyze; confront; prescribe; provide principles, models, and theories)

52. Information and concepts delivered by (measure and feedback; explicit demonstration; general discussion; explicit plans and models)

53. Point of change in firm's organization (top; middle; bottom)

54. Main client contact is (entrepreneur; corporate management; linkpin or one who relates different departments of firm or communications within entire organization; task manager; bureaucrat)

55. Type of management technology (organizing; planning; controlling; leading)

56. Type of management technology (purchasing; executive management; human resources; production/operations; marketing/distribution; financial management)

57. The client's firm can be characterized as an unstable organization. (I disagree strongly; I disagree; I neither disagree nor agree; I agree; I agree strongly)

58. Firm is (able to hire and retain highly trained personnel; able to hire and retain average skilled personnel; not able to hire and retain trained personnel)

59. Executive responsiveness in adapting to environmental conditions (highly adaptable; moderately adaptable; neither adaptable or unable to adapt; firm can adapt but inadequately; firm cannot adapt to changing environmental conditions)

IV. Local Environmental Factors

60. Market environment is (very uncompetitive; uncompetitive; neither competitive nor uncompetitive; competitive; very competitive)

61. Government regulations affect firm's operations (not at all; very little; some, but are not very important; greatly; very greatly)

62. Professional associations and universities support client management needs for primary training (most inadequately; inadequately; neither adequately nor inadequately; adequately; most adequately)

63. Professional associations and universities support client management needs for other training and assistance (most inadequately; inadequately; neither adequately nor inadequately; adequately; most adequately)

64. Physical infrastructure supports management (most inadequately; inadequately; neither adequately nor inadequately; adequately; most adequately)

65. The general environment in the country is unstable (I disagree strongly; I disagree; I neither disagree nor agree; I agree; I agree strongly)

66. Within the client's industry group, the environment is generally stable (I disagree strongly; I disagree; I neither agree nor disagree; I agree; I agree strongly)

FIELD RESEARCHER EVALUATION

Questions one through seventeen were answered according to the following:

1 = disagree strongly
2 = disagree
3 = neither disagree nor agree

4 = agree
5 = agree strongly
6 = not answerable

1. The initial task statement was accurate.

2. The task as articulated in the work plan was completed by the volunteer.

3. The accomplished project was valuable to the client firm.

4. Project overall can be considered a successful transfer of technology.

The task carried out changed positively some aspect of the management process of the client:

5. Organizing
6. Planning
7. Controlling
8. Leading
9. Technology search

10. The techniques transferred have been diffused within the company.

11. The techniques utilized have remained (stuck) in the firm.

12. The techniques transferred have been diffused outside of the firm.

13. The client should have sought further outside assistance to maintain the results of the project.

14. The client has sought further outside assistance to maintain the results of the project.

15. As a result of the project, the client has sought other outside managerial technology.

16. The client understood the value of the technology he was to receive before the project began.

17. The client gained an understanding of the value of managerial technology as a result of the project.

Questions eighteen through forty-nine were answered according to the following:

1 = very negative
2 = negative
3 = neutral
4 = positive
5 = very positive
6 = not answerable

The forces affecting potential change of the client firm at the time of the IESC project:

18. Market opportunity
19. External competitive threat
20. Internal crisis
21. Environmental turbulence
22. Government action
23. Source of supply (non-human)
24. External technology environment
25. Cost of technology
26. Appropriateness of technology

The internal factors affecting potential change of the client:

27. Internal technology (task)
28. Technology acquisition history
29. Organization structure
30. Presence of a positive organizational mystique
31. Charismatic leader
32. Management resources
33. General manpower resources
34. Ownership
35. Labor
36. Financial resources
37. Comprehension of technology value

Managerial Practices

38. Organizing
39. Planning
40. Controlling
41. Leading

The process of consultation factors which affected potential change from the specific project:

42. Volunteer's capabilities
43. Volunteer's strategy for consultation
44. Duration of consultation
45. Diagnosis of situation
46. Entry Stage of consultation
47. Implementation of consultation
48. Termination of consultation
49. Ongoing maintenance of change

CASE STUDY GUIDE

I. Identification

General history, ownership, products
Type of firm
Technology import history
Is unit part of other organization?
Size (employees, sales, management group)
Growth at time of project
Historical problems

II. Local Environment Factors

Risk environment
Government laws and policies
Market environment
Technological infrastructure
Physical infrastructure
Knowledge of environment by client
Change in industry
Dominant concerns from outside

III. Management Characteristics

- Task technology
- Organizing
  - Structure (description, hierarchy)
  - Policy (development, communication, hiring, promotion, hierarchy)
  - Blocks to technology search
  - Differentiation
- Leading
  - Decision-making (authority; policy-making, production, marketing; structure)
  - Trust, motivation, suggestions

Planning
  System (scope, type, time covered, use)
  Technology information search
Controlling
  Standard setting, scope of QC, cost controls, budget controls
Standard key decisions
Dominant internal concerns

IV. Process of Consultation

History of problems
Users (department, number, stages) and target of change (type and level)
Situation (crisis, etc.)
Strategy of change
Consultant background
Information delivery
Contact point
Type of technology (management process, function)
Results
  Management effectiveness
  Changes in planning, organizing, controlling, leading
Diffusion
  Economic impact
  Search behavior
  Culture of organization
  Related events (new licenses, products, etc.)
Missed possibilities
Forces working against change
Follow-up events to work for or against change
Conclusion (strategy and events worked together)

NOTES

1. Newman, William H.; Charles E. Summer; and Warren E. Kirby. *The Process of Management.* Englewood Cliffs, N.J.: Prentice-Hall, Inc., 1977.

2. Negandhi, A.R., and S.B. Prasad. *Comparative Management.* New York: Appleton Century Crofts, 1970.

3. Hornstein, Harvey A. and Noel M. Tichy. *Organization Diagnosis and Improvement Strategies.* Hopewell Junction, N.Y.: Behavioral Science Associates, 1973.

Blake, Robert R., and Jane Srygley Mouton. *Building a Dynamic Corporation Through Grid Organizational Development.* Reading, Mass.: Addison-Wesley Publishing Company, 1969.

Bennis, Warren; Kenneth Beune; Robert Chin; and Kenneth Carey. *The Planning of Change.* 3rd. ed. New York: Holt, Rinehart and Winston, Inc., 1976.

4. Ginsberg, Michael. "The Process Approach to Management Science Implementation." Doctoral Dissertation, Massachusetts Institute of Technology, 1975.

5. Terreberry, S. "The Evolution of Organizational Environments" *Administrative Science Quarterly*, 12, No. 4 (March 1968):590-613.

6. Svennilson, Ingvar. "Technical Assistance: The Transfer of Industrial Know-How to Non-Industrial Countries" *In Economic Development with Special Reference to East Asia*, K. Berill, ed. New York: St. Martin's Press, 1964.

Peno, Joseph. "The Multinational Firm" The Utilization and Development of Host Country High-Level Manpower." Ph.D. Dissertation, Michigan State University, 1975.

Behrman, Jack N., and Harvey W. Wallender. *Transfers of Manufacturing Technology Within Multinational Enterprises.* Cambridge, Mass.: Ballinger Publishing Company, 1976.

Bibliography

ARTICLES AND PRESENTATIONS

Askin, Peter W. "Intermediate Technology: An Informal Study." Prepared for the 18th Session of the Department of State Senior Seminar in Foreign Policy, 1975-1976.

Baranson, Jack. "The New Generation of Technology Transfer Agreements: Implications for U.S. Economic Strength." Presentation to President's Advisory Group in Contribution of Technology to Economic Strength. Washington, D.C., June 13, 1976.

Camp, H.H., and C.A. Rojas Magnon. "Recent Developments Under the Mexican Foreign Investment Law and the Law Regulating the Transfer of Technology." *Lawyer of the Americas* 8, no. 1 (1976).

Commission on Transnational Corporations (second session). "Conclusions and Recommendations of the Commission." Lima, Peru, March 1-12, 1976.

Davig, William A. "The Transfer of Production Technology to Brazil: A Study of Two Industrial Sectors." Program of Research on the Management of R&D. Evanston, Illinois: Northwestern University, 1974.

"El Impacto de las Inversiones Extranjeras Privadas Sobre el Progreso Cientifico y Technologico de America Latina." Second Reunion of CIEECC, February 8, 1971, Lima, Peru. Washington, D.C.: Organizacion de los Estados Americanos, 1971.

"Excerpts in the Final Report of CACTAL Related to Transfer of Technology." A Report of the Regional Scientific and Technological Development Program of the Department of Scientific Affairs in the Organization of American States. Washington, D.C.: Organization of American States, 1973.

"Experiment in International Transfer of Technology: A Pilot Project for Latin America." General Secretarial Report SG/P.1 EPTT/4. Published by the

Regional Scientific and Technological Development Program in the Department of Scientific Affairs. Washington, D.C.: Organization of American States, 1973.

Farmer, R.N., and B.M. Richman. "A Model for Research in Comparative Management." *California Management Review* 7, no. 2 (Winter 1964):55-68.

Ginsberg, Michael. "The Process Approach to Management Science Implementation." Doctoral dissertation, Massachusetts Institute of Technology, 1975.

Gluck, Frederick W., and Richard N. Foster. "Managing Technological Change: A Box of Cigars for Brad." *Harvard Business Review* 53, no. 5 (1975):139-49.

Goldring, Lionel. "The Management of Science and Technology." Lecture Series at New York University Graduate School of Business Administration, January-May 1976.

Gonzalez, R.F., and C. McMillan, Jr. "Universality of American Management Philosophy." *Journal of the Academy of Management* 4, no. 1 (April 1961):33-45.

Hymer, Stephen. "La Grande Corporation Multinationale." *Revue Economique* 19, no. 6 (November 1968):949-73.

_______. "The International Operations of National Firms: A Study of Direct Investment." Ph.D. Dissertation, Massachusetts Institute of Technology, 1960.

Kissinger, Henry. "UNCTAD IV: Expanding Cooperation for Global Economic Development." Speech at the Fourth Ministerial Meeting of the United Nations Conference on Trade and Development (UNCTAD) in Nairobi, Kenya. N.Y.: Bureau of Public Affairs, Office of Media Services, May 1976.

Koontz, H. "A Model for Analyzing the Universality and Transferability of Management." *Academy of Management Journal* 12, no. 4 (December 1969):415-30.

_______. "The Management Theory Jungle." *Journal of the Academy of Management* 4, no. 3 (December 1961):174-88.

Korea Institute of Science and Technology, Descriptive Brochure. (Seoul: KIST) No Date.

La Transferencia y Desarrollo de la Tecnologia. Bogota, Colombia: Comite de Accion Interamericana de Colombia, 1975.

Lima Declaration and Plan of Action on Industrial Development and Co-operation. Second General Conference of the United Nations Industrial Development Organization, Lima, Peru, 12-26 March 1975, pg. 5.

Lindblom, Charles E. "The Science of Muddling Through." In *The Progress of Management*, 2nd ed., Harold Lazarus, E. Kirby Warren and Jerome E. Schnee, eds. Englewood Cliffs, N.J.: Prentice Hall, Inc., 1972.

"Manila Declaration." *Transfer of Technology*, Section Five, 77/MM, III/49, 1974. New York: United Nations.

"Measures to Strengthen the Technological Capability of Developing Countries." Initiatives Resulting from the United Nations Seventh Special Session on Development and International Economic Cooperation, September 1-16, 1975.

Michaelis, Micael. "Technology for Business." White House Conference on the Industrial World Ahead: A Look at Business in 1990. Boston, Mass.: Arthur D. Little, Inc., 1972.

Mayo, Santos. "Interaction between Scientific-Technological and Productive Sectors: Viewpoints of Scientific Research and Education." In *Science, Government, and Industry for Development*, E. Ingerson and W. Bragg, eds. Austin: University of Texas Press, 1975.

Martinez Vidal, Carlos, Speech delivered at a meeting sponsored by the Council of the Americas (Washington, D.C.: February 26, 1979).

Nadler, David A., and Ed E. Lawler. "Motivation—A Diagnostic Approach." In *Readings on Managerial Behavior*, David A. Nadler and Michael L. Tuschman, eds. New York: Columbia University Press, 1976.

Nau, Henry R. "U.S. Policy for Technology and Development: Experience Problems, and Private Sector Role." In *Public Policy and Technology Transfer*, vol. IV, pp. 43-105. New York: Fund for Multinational Management Education, 1978.

Negandhi, A.R. "A Model for Analyzing Organizations in Cross-cultural Settings: A Conceptual Scheme and Some Research Findings." In *Comparative Administration and Research Conference*. Kent, Ohio: Kent State University, 1969.

Negandhi, A.R., and B.D. Estafen. "A Research Model to Determine the Applicability of American Management Know-How in Differing Cultures and/or Environments." *Academy of Management Journal* 8, no. 4 (December 1963):309-18.

Pavitt, Keith. "The Multinational Enterprise and the Transfer of Technology." In *The Multinational Enterprise*, John H. Dunning, ed. London: Allen and Unwin, 1971.

Peno, Joseph. "The Multinational Firm: The Utilization and Development of Host Country High-Level Manpower." Ph.D. Dissertation, Michigan State University, 1975.

_______. "Multinational Corporate Behavior in Host-Country High-Level Manpower Markets. The Implication for Technology Transfer and Foreign Investment Control in the Less Developed Host Country." O.E.C.D. Development Center Conference Paper Number 3—International Meeting of Research on Transfer of Technology by Multinational Firms, 1975.

Perrow, Charles. "A Framework for the Comparative Analysis of Organizations." *American Sociological Review* 32 (April 1967):194-208.

Prebisch, Raul. "Commercial Policies in the Underdeveloped Countries." *American Economic Review* 49, no. 2 (May 1959):251-73.

Sunkel, Oszaldo. "Big Business and 'Dependencia,' A Latin American View." *Foreign Affairs*, April 1972, pgs. 517-531.

Svennilson, Ingvar. "Technical Assistance: The Transfer of Industrial Know-how to Non-Industrial Countries." In *Economic Development with Special Reference to East Asia*, K. Berill, ed. New York: St. Martin's Press, 1964.

Quinn, James Brian. "Technological Forecasting." *Harvard Business Review* 45, no. 2 (March-April 1967):89-105.

"Resumen de los Criterios Generales de Aplicacion de Ley sobre el Registro de la Transferencia de Tecnologia y el Uso y Explotacion de Patentes y Marcas." Mexico, D.F.: Government of Mexico, 1974.

Rosas, Gabriel V. "Algunas Consideraciones sobre Transferencia de Tecnologia." *Boletin del CICYP* (Bogota, Columbia) 3a Epoca, no. 16 (Enero 1976), pg. 12.

Rubenstein, Albert H. "Technical Information, Technical Assistance, and Technology Transfer—The Need for a Synthesis." Presented at the seminar on the Management of the Transfer of Technology within Industrial Cooperation. Geneva: Economic Commission for Europe, United Nations, July 14, 1975.

Sagasti, Francisco. "A Framework for the Formulation and Implementation of Technology Policies: A Case Study of ITINTEC in Peru." In *Science, Government and Industry for Development*, Ingerson, E. and W. Bragg, eds., pp. 203-244. Austin: University of Texas Press, 1975.

Schydlowsky, Daniel M., Wicht, Juan J., "The Anatomy of an Economic Failure: Peru 1968-78," *The Peruvian Experiment Revisited*, ed. A. Lowenthol (Boston: Center for L.A. Studies, Boston University, 1979).

Terreberry, S. "The Evolution of Organizational Environments." *Administrative Science Quarterly* 12, no. 4 (March 1968):590-613.

Truitt, Nancy S. "Assessing the Technology Transferred to Developing Countries by Multinational Corporations." Paper submitted to Professor Lionel Goldring of New York University, May 24, 1976.

UNCTAD. "The Transfer of Technology." *Journal of World Trade Law* 4 (September-October 1970):711.

Varsavsky, Carlos M. "On the Inward and Outward Looking Components of the Scientific Community." In *Views on Science, Technology, and Development*, Eugene and Victor Rabinowitch, eds. Oxford, England: World Development Publishing Company, 1976.

Vertinsky, Ilan. "OR/MS Implementation in Valle, Colombia, S.A.—A Profile of a Developing Region." *Management Science* 18, no. 6 (February 19, 1972), pgs. B314-B327.

Vertinsky, Ilan, and Richard T. Barth. "A Model for Diffusion and Implementation: An Exploratory Study of Managerial Innovation in Colombia." In *Social Economic Planning Science*, vol. 6, pp. 150-71. London: Pergamin Press, 1972.

Wallender, Harvey W. "Sources of Conflict Associated with International Transfer and Development of Technology." Presented at the Quadrilateral Businessmen's Conference on Economic Interdependence, October 29, 1977, London. Washington, D.C.: U.S. Chamber of Commerce, 1977.

_______. "Managing and Evaluating Technology Transfer in Manufacturing Industries." Presented at the seminar on the management of transfer of technology within industrial cooperation sponsored by the Economic Commission for Europe of the United Nations. Document Identification: SC, TECH., SEM 3/R.16. Geneva: United Nations Economic and Social Council, 1975.

Walters, Ingo and Thomas N. Gladwin, "Technology and Conflict in Multinational Corporation Operations," *Multinationals in Conflict: Lessons in Conflict Management* (New York: John Wiley and Sons, forthcoming late 1979).

Wionczek, M.S. "A Latin American View." In *How Latin America Views the U.S. Investor*, Vernon, R., ed. New York: Praeger Publishers, 1966.

Zand, Dale E., and Richard E. Sorensen. "Theory of Change and the Effective Use of Management Science." *Administrative Science Quarterly* 20 (December 1975), pgs. 532-545.

BOOKS

Adams, Scott. *Scientific and Technical Information Services in Eight Latin American Countries.* Washington, D.C.: National Science Foundation, 1975.

Analytical Methods in Government Science Policy and Evaluation. Paris: OECD, 1972.

Ayala, Hector. *Transfererencia de Tecnologia de Companias Multinacionales a Paises en Desarrollo.* Casos en la Industria Colombiana. Bogota: CLADEA and Fund for Multinational Management Education, 1975.

Barnard, Chester I. *The Function of the Executive.* Cambridge, Mass.: Harvard University Press, 1968.

Behrman, Jack N., and Harvey W. Wallender. *Transfers of Manufacturing Technology Within Multinational Enterprises.* Cambridge, Mass.: Ballinger Publishing Company, 1976.

Bennis, Warren; Kenneth Beune; Robert Chin; and Kenneth Carey. *The Planning of Change.* 3rd ed. New York: Holt, Rinehart and Winston, Inc., 1976.

Blake, Robert R., and Jane Srygley Mouton. *Building a Dynamic Corporation Through Grid Organizational Development.* Reading, Mass.: Addison-Wesley Publishing Company, 1969.

Burns, Tom, and G.M. Stalker. *Management of Innovation.* 2nd ed. London: Tavistock Press, 1966.

Chapple, E.C., and L.R. Sayles. *The Measurement of Management.* New York: MacMillan, 1961.

Choice and Adaptation of Technology in Developing Countries. Paris: OECD Development Centre, 1974.

Codes of Conduct for the Transfer of Technology: A Critique. New York: Fund for Multinational Management Education, 1976.

Cooper, Charles, ed. *Science, Technology and Development.* London: Frank Cass, 1973.

De Cubas, Jose. *Technology Transfer and the Developing Nations.* New York: Fund for Multinational Management Education, 1974.

Driscoll, Robert E., and Harvey W. Wallender. *Technology Transfer and Development: A Historical and Geographic Perspective.* New York: Fund for Multinational Management Education and the Council of the Americas, 1974.

Farmer, R.N., and B.M. Richman. *Comparative Management and Economic Progress.* Homewood, Ill.: Richard D. Irwin, Inc., 1966.

Fayol, Henri. *Industrial and General Administration.* Translated by J.A. Coubrough. Geneva: International Management Institute, 1929; and Henri Fayol, *General and Industrial Management.* London: Sir Isaac Litman and Sons, Ltd., 1949.

French, Wendell L., and Cecil H. Bell. *Organization Development.* Englewood Cliffs, N.J.: Prentice-Hall, Inc., 1973.

Haire, M.E.; F. Ghiselli; and L.W. Porter. *Managerial Thinking: An International Study.* New York: John Wiley and Sons, Inc. 1966.

Hampton, David R.; Charles E. Summer; and Ross A. Webber. *Organizational Behavior and the Practice of Management.* Glenview, Ill.: Scott, Foresman and Company, 1973.

Harbison, F., and C.A. Myers, *Management in the Industrial World.* New York: McGraw-Hill Book Company, 1959.

Hicks, Herbert G. *The Management of Organizations: A System and Human Resources Approach.* New York: McGraw-Hill, Inc., 1972.

Hornstein, Harvey A., and Noel M. Tichy. *Organizational Diagnosis and Improvement Strategies.* Hopewell Junction, N.Y.: Behavioral Science Associates, 1973.

Hornstein, Harvey A.; Barbara B. Bunker; Warner W. Burke; Marion Gindes; and Roy J. Lewicki. *Social Intervention, A Behavioral Science Approach.* New York: The Free Press, 1971.

Ingerson, Earl, and Wayne G. Bragg, eds. *Science, Government, and Industry for Development.* Austin: University of Texas Press, 1976.

Jequier, Nicholas. *Appropriate Technology, Problems and Promises.* Paris: OECD Development Centre, 1976.

Katz, Daniel, and Robert L. Kahn. *The Social Psychology of Organizations.* New York: John Wiley and Sons, 1966.

Katz, Jorge M. *Importacion de Tecnologia, Aprendizaje Local e Industrializacion Dependiente.* Buenos Aires: Instituto Torcauto de Tella, 1972.

Kolb, David A., and Alan L. Frohman, "An Organization Development Approach to Consulting," *Sloan Management Review* vol. 12, no. 1 (1970):51-65.

Koontz, H. *Toward a Unified Theory of Management.* New York: McGraw-Hill Book Company, 1964.

Lawrence, Paul R., and Jay W. Lorsch. *Organization and Environment.* Homewood, Ill.: Richard D. Irwin, Inc., 1966.

Likert, Renis. *New Patterns of Management.* New York: McGraw-Hill Book Company, 1961.

Lindzey, G., and A. Aronson. *The Handbook of Social Psychology.* Reading, Mass.: Addison-Wesley Publishing Company, 1968.

Lippitt, Ronald; Jeanne Watson; and Bruce Westley. *The Dynamics of Planned Change.* New York: Harcourt, Brace and World, 1958.

Machiavelli, Nicolo. *The Prince.* London: Oxford University Press, 1968, p. 24.

Manning, G.K. *Technology Transfer: Successes and Failures.* San Francisco: San Francisco Press, 1974.

March, James J., and Herbert A. Simon. *Organizations.* New York: John Wiley and Sons, Inc., 1958.

Michigan Organizational Assessment Package. Ann Arbor: University of Michigan, 1975.

Mintzberg, Henry. *The Nature of Managerial Work.* New York: Harper and Row, 1973.

Newell, Allen, and Herbert A. Simon. *Human Problem Solving.* Englewood Cliffs, N.J.: Prentice-Hall, 1972.

Newman, William H.; Charles E. Summer; and Warren E. Kirby. *The Process of Management.* Englewood Cliffs, N.J.: Prentice-Hall, Inc., 1977.

Negandhi, A.R., and S.B. Prasad. *Comparative Management.* New York: Appleton Century Crofts, 1970.

Nie, Norman H., and Hull C. Hadlai. *Statistical Package for the Social Sciences—SPSS.* 2nd ed. New York: McGraw-Hill, Inc., 1975.

Okolie, Charles Chukwuma. *Legal Aspects of the International Transfer of Technology to Developing Countries.* New York: Praeger Publishers, 1975.

Proposal for a Program in Appropriate Technology. Report of the Agency for International Development to the United States Congress pursuant to the Section 197 of the Foreign Assistance Act. Washington, D.C.: U.S. Government Printing Office, 1976.

Public Policy and Technology Transfer, vol. I-IV. New York: Fund for Multinational Management Education, 1978.

Richmond, Samuel B. *Operations Research for Management Decisions.* New York: Ronald Press Company, 1968.

Runkel, P.J., and Joseph E. McGrath. *Research on Human Behavior.* New York: Holt, Rinehart and Winston, Inc., 1972.

Saint-Rossy, D.T.; A. Chakrabarti; and A.H. Rubenstein. *International Transfer of Technology—An Evaluation of the Education and Training Component.* Evanston, Ill.: The Technological Institute of Northwestern University, May 1976.

Schein, Edgar H. *Process Consultation: Its Role in Organization Development.* Reading, Mass.: Addison-Wesley Publishing Company, 1969.

Strassmann, Paul W. *Technological Change and Economic Development.* Ithaca, N.Y.: Cornell University Press, 1968.

Summary and Update: Foundation for International Technological Cooperation. Washington, D.C.: Planning Office, October 31, 1978.

Taylor, Frederick. *The Principles of Scientific Management.* New York: Harper and Brothers, 1911.

Teece, David. *The Multinational Corporation and the Resource Cost of International Technology Transfer.* Cambridge, Mass.: Ballinger, 1976.

U.S. International Firms and R D and E in Developing Countries. Washington, D.C.: National Academy of Sciences, 1973.

U.S. Science and Technology for Development: A Contribution to the U.N. Conference. Prepared by National Research Council, National Academy of Science, National Academy of Engineering, and Institute of Medicine. Washington, D.C.: Department of State, 1978.

Vaitsos, Constantine V. *Intercountry Income Distribution and Transnational Enterprises.* Oxford: Clarendon Press, 1974.

Weber, Max. *Social and Economic Organization.* New York: MacMillan Company, 1947.

Woodward, Joan. *Management and Technology.* London: Her Majesty's Stationery Office, 1958.

Wortmann, Max W., and Fred Luthans. *Emerging Concepts in Management.* 2nd ed. New York: MacMillan, 1975.

Zimbardo, Philip, and Ebbe Ebbeson. *Influencing Attitudes and Changing Behavior.* Reading, Mass.: Addison-Wesley Publishing Company, 1970.

Index

About the Author

Harvey W. Wallender III is Managing Director for the Council of the Americas and Director of Planning for the Fund for Multinational Management Education. Since 1974, he has also acted as a consultant for a number of corporations and the U.S. Department of State in the areas of strategic planning for international operations and transfer of technology.

Professor Wallender is a member of the Management and International Business faculty at the Pace University Graduate School of Business and a visiting lecturer at the Center for Transnational Management Development at George Washington University.

Mr. Wallender was formerly Director of Planning and New Products for the Financial Services Group of the International Basic Economy Corporation (IBEC) and has acted as President of the Transoceanic Fishing Corporation and Director of Planning for the Food Products Division of IBEC.

Mr. Wallender's undergraduate and graduate work in political science was done at the University of Texas followed by graduate research at the University of Madrid. He also holds an M.B.A. from Columbia University and was a Ph.D. candidate in the Department of Management.

Mr. Wallender has published a wide variety of articles and research monographs on strategic planning, Latin American business environment, and the international transfer of technology. He co-authored with Professor Jack N. Behrman *Transfers of Manufacturing Technology Within Multinational Enterprises*, and was the Director of the research program that has developed a four volume series entitled *Public Policy and Technology Transfer: Viewpoints of U.S. Business* for the U.S. Department of State.